# Wasting the Rain

First published in 1992, this title offers an experienced and constructive evaluation of the ways in which water resources have been developed in Africa. Adams argues that the best hope of productive development lies in working and engaging with local people and using local knowledge of the environment effectively. Modern, large-scale developments that have largely been ineffective are examined, and emphasis is placed on the importance of using the skills and concerns of those affected, such as small farmers, to develop ingenious water projects – an approach that can be applied worldwide. This is an interesting and relevant title, which will be of particular value to those with an interest in the developments in water resource conservation over the past two decades.

# Wasting the Rain

Rivers, People and Planning in Africa

## W. M. Adams

Routledge
Taylor & Francis Group

First published in 1992
by Earthscan Publications Ltd

This edition first published in 2013 by Routledge
2 Park Square, Milton Park, Abingdon, Oxon, OX14 4RN

Simultaneously published in the USA and Canada
by Routledge
711 Third Avenue, New York, NY 10017

*Routledge is an imprint of the Taylor & Francis Group, an informa business*

© 1992 W. M. Adams

**Publisher's Note**
The publisher has gone to great lengths to ensure the quality of this reprint but points out that some imperfections in the original copies may be apparent.

**Disclaimer**
The publisher has made every effort to trace copyright holders and welcomes correspondence from those they have been unable to contact.

A Library of Congress record exists under LC control number: 92022825

ISBN 13: 978-0-415-71872-1 (hbk)
ISBN 13: 978-1-315-86786-1 (ebk)
ISBN 13: 978-0-415-71880-6 (pbk)

# Wasting the rain

Rivers, people and planning
in Africa

W. M. Adams

EARTHSCAN
Earthscan Publications Ltd, London

First published 1992 by
Earthscan Publications Ltd
120 Pentoville Road, London N1 9JN

# Contents

# List of figures and tables

## Figures

## Tables

# Preface

Africa's river valleys have been important targets for rural development planners. Most major African rivers have been dammed, some several times, and irrigation schemes have been built in many floodplains. In some cases the purpose of such development has been to increase production of agricultural crops, to tackle disease or problems of basic human needs. Other projects have been designed to meet urban needs or the demands of the industrial economy, for example the construction of dams for urban water supply or for the generation of hydro-electric power.

There have been some successes among these schemes, but also many failures. Their planners, of course, have mostly gone home. The projects remain behind, and their benefits are enjoyed – or their impacts are endured – by the rural people who were caught up in their development. Seasonal river flows in arid and drought-prone savanna environments have frequently suggested the possibility of irrigation and permanent cultivation. The sight of floodplains and swamps where water is 'wasted' in providing fishing, grazing and simple flood-agriculture has spurred economists and agriculturalists to try to control floods and to build irrigation schemes to try to increase agricultural production. The sight of rivers flowing between hills has been an invitation to engineers to build dams for power generation.

Such projects have been only one of the many forms of physical, economic, institutional and cultural interventions by outsiders in African rural environments and economies. Most of these interventions follow established planning procedures, although sophisticated techniques of project appraisal are relatively recent. These techniques make an attempt to estimate the future benefits and future costs of project development, or to assess the relative merits of different alternatives. Many of these projects have been funded by First World banks, bilateral or multilateral aid donors, and have been subject to the increasingly stringent financial, economic and (recently at least) environmental assessment procedures of these agencies. Nonetheless, the success rate has been poor. Despite these failures, the enthusiasm of African governments and their advisers and bankers for large-scale projects to transform the African environment persists.

It is this intervention to develop Africa's water resources, and the persistence of the mentality that gives rise to grand schemes and sweeping environmental transformation, that form the subject of this book.

Mistakes are common in development in Africa, and this book tries to analyse them. It seeks to do so constructively, and in a spirit of shared learning. My own experience as an outsider in Africa, first as a consultant and latterly as a researcher, have left me only too aware of the limitations of my own knowledge and understanding, both of the world of development professionals and (all the more) of Africa itself.

Where possible the book is written from first-hand experience, and for that reason many of its examples are drawn from those places with which I am familiar. As a result the regional coverage is far from perfect, and will not please everyone, although it is my belief that the arguments put forward will be true of other places. There is, of course, an arrogance in any outsider writing a book about African development, and a peculiar irony when that book contains a critique of the work of other outsiders and the changes they bring. In reply I would only say that I have written what I believe to be true, and see no reason not to do so in a spirit of open debate. Much well-meaning effort has been expended in Africa, and many bold schemes have failed. I am convinced that it is important to be honest about that failure if we are to understand what has been done in the past, and start to imagine what action might be needed now to support Africa's floodplain peoples in their search for a sustainable future.

W. M. Adams
Department of Geography
University of Cambridge
May 1992

# Acknowledgements

This book was written while I was on leave from the University of Cambridge at the University of Calgary against the background of controversy in Canada about the construction and environmental appraisal of the Oldman and Rafferty-Alameda Dams in Alberta and Saskatchewan and the James Bay II (Great Whale) hydro project in Quebec. The issues involved would be very familiar to many riverine people in Africa.

I would like to thank the Department of Geography and Downing College in Cambridge for letting me step off the treadmill, and the Faculty of Environmental Design at the University of Calgary for making me so welcome and providing me with an excellent environment to work in. I am particularly grateful to Philip Elder for letting me occupy his desk, and for setting up my visit. I would like to thank Nigel and Susan Waters, Jeff Gruttz, Whiskey Jack, and other friends for keeping me sane, and Emily and Thomas for many pleasurable diversions from the task in hand. I would like to thank Franc for reading through the chapters as they appeared and making encouraging noises while providing an effective critique of the contents.

The maps and diagrams were drawn in the Department of Geography in Cambridge by Michael Young and Ian Gulley. Figures 8 and 18 were previously published in *World Development* Volume 18 (Pergamon Press) and Figure 19 appeared in the *Transactions of the Institute of British Geographers* Volume 16 (1991). Jane Robinson helped valiantly with proof-reading.

A great many people have contributed to my education over the fourteen years or so that I have been working on African water resource development, either in a formal sense or through shared work. It would be impossible to thank them all. A.T. Grove, Alan Bird and Jack Griffith got me started. More recently, Martin Adams, Richard Carter, Pat Dugan, Ted Hollis, Kevin Kimmage and David Thomas have had a particular influence on my thinking. I would like to thank Pat Dugan, Martin Adams, Geoff Howard and Paul Richards for their comments on some or all of the manuscript. I have used their ideas freely, although (of course) responsibility for what I have written here remains my own.

# CHAPTER ONE

# Introduction

*For who hath despised the day of small things?*[1]

## Wasting the rain

My first experience of Africa was life on a large engineering project in Nigeria at the end of the 1970s. A dam and irrigation scheme were being built along a river valley in dry savanna country. This area lay south of the infamous Sahel zone, but it shared some of its characteristics. It had a short rainy season, it suffered from periodic drought and had poor soils. In the light of this litany of problems (all of them widespread in Africa) it is perhaps surprising that the valley should have been already quite intensively farmed, but it was so. It was densely populated, and land was in permanent use. The irrigation scheme was going to transform the existing patterns of farming in the valley, and with it the lives of the valley's people. In place of reliance on rain and natural river flooding, the dam would supply plentiful water for irrigation throughout the year. Crops could be grown in the dry season, new services and agricultural supplies like fertilizers could be supplied, and yields would be higher and more reliable. The farmers would make more money, the country would have more food, and the constant threat of drought and hunger would recede. It was a generous and exciting vision.

Early in my experience of the project, I asked a young engineer what he thought of the idea of damming the river and intensifying agriculture through irrigation. His reply was simple, his conviction complete: 'It is the only way to stop the water running to waste'. To him, the annual floods in the wet season were wasteful, and the farmers' attempts to grow crops using rainfall and river floods were rather pathetic. He had seen the farmers from his truck window struggling with poverty and a harsh environment. They were trapped by nature, without the knowledge, the skills and the resources to break out. The construction camp where he lived had a swimming pool, a supermarket and air conditioners, glaring reminders of the wealth of developed Western countries from which the construction engineers came. The technical sophistication of the construction processes and the shattering difference in wealth, outlook and resources between the builders and the farmers around them confirmed for my young engineer the rightness of the task in hand. Only a major transformation of environment, society and economy could change conditions on the ground. Only the boldest, most modern and imaginative developments could counteract the poverty and backwardness of the farmers and the harshness of their environment. Only modern technology could stop the rain from running to waste.

Unfortunately, the reality has been rather different. A decade on, the scheme lay partly disused. Many farmers had fled, others worked as wage-labourers on land they once controlled. Machines lay idle and deteriorating, while canals had silted up and drains eroded. Downstream, the floodplain received less water and the area of dry season agriculture had shrunk. What had gone wrong? This is an obvious question, and it is one to which commentators on Nigerian development have given much attention. Was it an isolated failure of project planning, an unfortunate blot on an otherwise reasonable record of success? Or was it typical of other projects, other failures?

I do not know where my engineer friend is today. Probably he is in some remote corner of the globe building another irrigation scheme, while I sit at my word processor and rehash a conversation he has surely long forgotten. But if he were here, I would ask him one question: Who was wasting the rain? Was it the farmers of the valley, or the multi-million pound irrigation scheme? With the benefits of hindsight, he would probably now agree that despite appearances the farmer was doing pretty well in difficult conditions. Despite its high-tech and grand plans, the irrigation scheme has turned out to be a terrible waste of money, human resources and . . . water.

Yet that irrigation scheme is not unusual. It is certainly not unique. It grew out of a particular set of ideas about what could be done to develop the people and environments of sub-Saharan Africa, and what should be done. Africa bristles with similar projects, dams on major rivers and canals through the savanna. Many of them are in disarray, some are abandoned. Others are finished and functioning producing hydro-electric power or irrigated crops as they were planned to, but at a high economic cost. Many water resource projects in Africa have failed. The World Bank accepts the results of a 1987 study that shows that half the rural development projects they have funded in sub-Saharan Africa have failed. It comments 'there are countless examples of badly chosen and poorly designed public investments . . . African governments and foreign financiers (commercial banks and export credit agencies as well as donor agencies) must share responsibility'.[2] The World Bank is by no means alone in having projects that perform badly. The record of many aid agencies and governments in African development is remarkably poor. Many projects have failed to bring a sustained fall in poverty, any enhancement of the environment and the things it can yield to the poor, or an increase in the power of the poor to help themselves. This kind of 'development' is both expensive and destructive. In economies short of capital, skills, resources and time, it is disastrous.

These projects are all products of a vision of Africa developed and transformed. The vision sees positive change coming from technology, a modernized market economy, a community of both small and large businessmen-farmers. But is such development all gain? Denis Goulet calls development 'an ambiguous process', commenting that it is often 'depicted as a crucible through which all societies must pass and, if successful, emerge purified: modern, affluent, and efficient. But such a portrait is misleading: it confuses a part of contemporary history with the larger whole'.[3] In reality, development projects create losers (at least in the short term) as well as winners. Project development also carries risks, and has opportunity costs, taking money that might be better spent on something else.

Not only are the ambiguities of the changes in this 'development' often ignored, but the process itself is very often seen to be something that can only be begun from outside, from outside Africa, and from outside rural society. This dominant view of 'development' is that it is something to be done *for* people (and to them), not something people do by themselves. Although many people and agencies have moved beyond this view, it is still widely held, even by those who consider themselves 'experts' in development, entrenched in the structures and established thinking of institutions and agencies of development. In pursuit of this idea of development billions of dollars have been spent in recent decades, millions of hectares of land have been developed and millions of lives have been changed, very often to little good effect.

This vision of development as something that outsiders can do that will dramatically transform Africa is bold and optimistic, and it sets a demanding agenda for action. It requires that natural resources be developed to meet human needs using the biggest and best technology available. It assumes that if enough money is invested in the right way, the poor countries of sub-Saharan Africa can become self-sufficient; their economies can run with those of the developed First World, and their people can attain the standards of living and life expectancy that exist in Europe or North America. The vision is cornucopian. The horn of plenty is ready to flow with good things, if only we get the planning right. Nature can be transformed, its wealth harnessed and serving humankind, and expert planning can handle the environmental and human implications of change. Development means stopping that wealth running to waste: no longer wasting the rain.

This is a grand vision. Unfortunately it is flawed. In order to understand why, we must look both at its origins – the ideas and people from which it comes – and its impact on the land and water of sub-Saharan Africa. Furthermore, we must look for alternatives. An increasing number of people in the development business are starting to challenge the accepted notions of this vision of development. What should replace it? What is needed is a new approach to development, combining integrated natural resource management with realistic socio-economic goals. The aim of this book is to build on an analysis of past development to explore the feasibility of new approaches to the rivers and wetlands of Africa.

## African rivers and wetlands

The Africa discussed in this book is 'sub-Saharan' Africa. It covers some 21 million square kilometres, and includes 46 countries. These range from the vast areas of the Sudan, Chad or Zaire to the tiny countries of Burundi, Gambia or Equatorial Guinea (see Figure 1.1). It excludes those countries abutting the Mediterranean (Egypt and the Magreb; Libya, Tunisia, Algeria and Morocco) and the Republic of South Africa whose geographies (for very different reasons) are so distinct from that of the tropical African countries that they require books to themselves.

The equator runs through the middle of sub-Saharan Africa (through Kenya, Uganda and Zaire), and most of the continent lies between the two tropics. Its climates range from the semi-desert of the southern fringe of the Sahara (in Mauritania, Mali or Sudan) or the edges of the Kalahari (Botswana), though a wide diversity of arid grasslands and thorn-scrub savannas to the lush rainforests of coastal West Africa (Ivory Coast, Liberia or Cameroon, for example) and Central Africa (Zaire, Republic of Congo and Gabon). There are mountain environments also, notably perhaps in the highlands of Ethiopia, East Africa (Kenya, Tanzania and Rwanda/Zaire), and in isolated spots in West Africa. There are also a wide range of wetland environments.

Wetlands are 'ecosystems whose formation has been dominated by water, and whose processes and characteristics are largely controlled by water'.[4] There is now growing international concern about wetlands

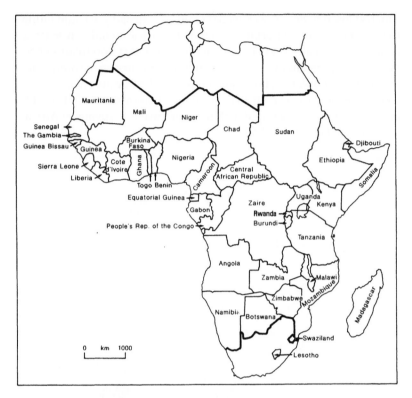

**Figure 1.1** The countries of sub-Saharan Africa

because of their enormous economic importance, their sensitivity to ecological change and their attractiveness for other land users such as intensive agriculture and industrialization.[5] They comprise about 6 per cent of the land area of the earth.[6] Engineers, hydrologists and economists concerned with development planning tend not to use the term 'wetland', but to refer to rivers, lakes or swamps in their proposals, or speak of river basins (meaning the whole hydrological system from source downwards). However, as the importance of the integration and land and water resources is recognized, the concept of wetlands is being increasingly widely adopted.

Wetland environments began to receive widespread attention in the Third World following signing of the Ramsar Convention in 1971. This defined wetlands very broadly to include the widest possible variety of ecosystems. The definition is worth quoting in full. Wetlands were defined as 'areas of marsh, fen, peatland or water, whether natural or artificial, permanent or temporary, with water that is static or flowing, fresh, brackish or salt, including areas of marine water the depth of

which at low tide does not exceed six metres'.[7] Broadly, it could be said
that economically important wetlands in Africa include river floodplains (on all scales from 1–2 km across to the vast floodplains of the
major rivers), freshwater swamps (sometimes occurring in conjunction
with these floodplains), lakes, and coastal and estuarine environments
(including mangroves). Of all the wetland environments of Africa, this
book is primarily concerned with the extensive floodplains of Africa's
major rivers (see Figure 1.2).

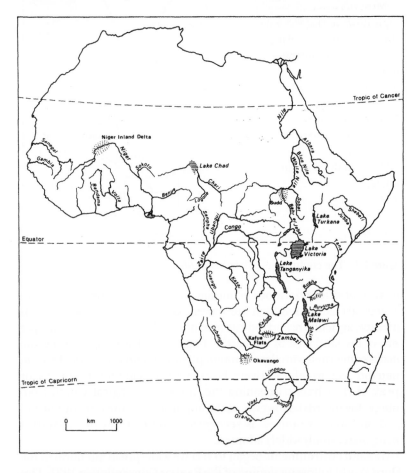

**Figure 1.2**  Major rivers and wetlands of sub-Saharan Africa

There are fringing floodplains alongside river channels on most Afri-
can rivers, for example the Senegal and Niger in West Africa or the
Zambezi, Rufiji, Tana and Jubba Rivers in East Africa, or on the Nile.
Although it is so massive, the River Zaire has very little seasonal flood-

plain because its flow is relatively unseasonal, although there are large areas of forested swamp. Larger floodplain areas, sometimes called internal deltas, occur on the Niger in Mali, the Kafue in Zambia and on the Chari-Logone system in Cameroon. These are often linked to permanent swamp systems – for example in the Sudd in the Sudan, the Okavango Delta in Botswana and on the Shire River draining Lake Malawi. There are also very coastal deltaic floodplains on a number of rivers, for example the Senegal, Niger and Zambezi.

Most of Africa's rivers have strongly seasonal flooding patterns, with high flows in the wet season and extensive flooding, and low flows in the dry season. During the flood period, water overflows the river banks, causing widespread inundation of the floodplain. The area inundated then falls with the river hydrograph, until in the dry season some rivers can be reduced to pools of water separated by dry land. In the Senegal valley some 5,000 km$^2$ is flooded at high flow, and about 500 km$^2$ in the dry season. The inundated area in the fringing floodplain of the Niger covers about 6,000 km$^2$ in the flood season, shrinking to about half that at low water. The Niger Inland Delta extends to 20,000–30,000 km$^2$ in the flood season, shrinking to 4,000 km$^2$ at low water. In the Kafue Flats in Zambia the area flooded varies from 28,000 km$^2$ in the wet season to 13,000 km$^2$ in the dry season. In the complex Logone-Chari floodplains south of Lake Chad (the Yaérés), flooding covers some 90,000 km$^2$, of which only 7 per cent remains wet at low water.

Floodplains consist of a complex of physical features left by past deposition and erosion by river channels. There are often levées, areas of relatively high land near the river channel which are formed of coarse sediment dropped as the river overflows its banks, and lower-lying areas of fine clays and silts further away from the river. The range of soil and flooding conditions even within a small area of floodplain can therefore be very great. Some parts of the floodplain will be inundated for only a very short period, and perhaps not at all in low-flood years, while other areas flood for many months, or perhaps form permanent water bodies.

The timing and duration of flooding changes downstream in a floodplain river. In the upper reaches of a river, where floodplains tend to be narrower, rivers tend to have a quick response to rainfall so that floods tend to have a sharper peak and shorter duration. Further downstream, where floodplains are larger and have a more complex form, flood peaks are delayed and less sharp. Thus as a flood peak moves down a river such as the Benue in Nigeria or the Tana in Kenya, the peak of the flood becomes progressively lower, and its rate of advance is delayed so that it is slower to begin, it lasts longer, and hence is later to end. On the Niger River the flood peak moves downstream at about 17 km per day, taking over 100 days to travel the 1,760 km from Koulikoro to Malanville.[8]

Flooding does not necessarily all come from river flow, however. Local rainfall can be important, particularly in flooding backswamp areas and pools. In some cases local runoff can cause flooding before river inundation begins, which can be important for the timing of agricultural or grazing activity. In deltaic environments tidal movements can be important, backing up river flows and enhancing freshwater flooding. This is a significant factor in the flood-related agriculture of the Senegal floodplain and the Basse Casamance in West Africa. Local runoff into the floodplain also provides a vital source of water for irrigation in areas such as the Basse Casamance which are subject to brackish or salt tidal influence.

Despite the aridity of much of Africa, the rivers and the extensive flooded wetlands around them support human communities engaged in agriculture, fishing or livestock husbandry. Smaller wetlands in dry areas such as the wadis of Kordofan or the Red Sea Province of Sudan, or the dambos of Zambia or Zimbabwe, are important to people in similar ways.[9] Wetlands provide a range of resources for human use, and provide what are effectively 'free' economic benefits. These are often referred to as the 'functions' of the wetland. They result from the interconnection of the geomorphological, hydrological and ecological processes going on in the wetland area. For example, where floodwater from the River Niger floods out from the river channel to fill the channels and pools of the Niger Inland Delta in Mali, a substantial amount of water percolates down to groundwater. Flooding within a wetland can sustain groundwater levels in aquifers many kilometres away from the wetland itself. Wetlands also provide a range of goods, such as water supplies, forage and hunting resources, wood resources, grazing, fish and agricultural produce. Furthermore, a single wetland may produce a number of these outputs at the same time, or serve different communities in different ways through a year.[10]

In dryland Africa, the economic importance of wetlands is very great. Sometimes this fact is lost amidst concern about the development problems and needs of drylands, and visions of huge benefits somewhere in the future if wetlands are 'developed'. However, even without such development, wetland areas have an important place in the economy of many African countries. This can include direct production of surplus food or other commodities or simply providing sound and sustainable incomes in both good and bad years for fairly large numbers of people. IUCN research on the the Niger Inland Delta, for example, shows that it supports some 550,000 people, and in the dry season provides grazing for about 1–1.5 million cattle, 1 million sheep and goats and 0.7m camels. There are some 80,000 fishermen, and the Delta supports extensive areas of rice (see Chapter 4).[11]

## Wetlands and development in africa

There are dams on almost all major African rivers, and many have been the targets of other kinds of development also, particularly irrigation schemes. Indeed, Africa's wetlands have been constant sources of inspiration for would-be developers. The geologist Albert Kitson, drifting down the Volta River in a canoe in 1915, saw the falls of Akosombo and dreamed of a dam to produce hydro-electricity to smelt the bauxite and produce aluminium. Kitson's dream lay fallow for many years, but parts of it were realized when construction of the Akosombo Dam on the Volta began in 1961.

Unfortunately, as in so many other cases, the development has proved less successful than anticipated. The Volta River Project sold power at fixed low rates to a multinational aluminium company, which used imported and not local bauxite. There has been very little trickle-down economic benefit, no real 'development'.[12] Furthermore, in the early 1980s drought reduced the discharge in the River Volta and interrupted power generation at Akosombo. For a time the aluminium smelter at Tema was forced to close altogether.[13]

It is this contrast between the indigenous economies of African floodplains and the impacts of development which form the subject of this book. Chapter 2 looks at the ways in which outsiders have viewed and understood Africa, and the impacts their interventions have had. It discusses the development of African economies and the changing patterns of African agriculture. Too many outside development professionals have been oblivious to the evidence of their eyes, and have failed to recognise the skills and experience of local farmers, fishermen and pastoralists. Chapter 2 also discusses the nature of the 'indigenous technical knowledge' of African rural people, and argues that while without doubt some people fall into the trap of viewing rural Africa through rose-tinted spectacles, indigenous knowledge and indigenous skills and organisation are vital for a sustainable future for Africa.

Chapter 3 presents some basic information about Africa's environment, its rainfall and its rivers. It discusses the ways in which African rural people use the dryland environments, and particularly the ways in which pastoralism and agriculture are adapted to drought and rainfall variability. It also looks at the difficulties that face those who try to make technical predictions about future hydrological patterns. The African environment is not easily understood, and many projects have suffered because of inadequate technical analysis or inadequate data. The problems that Africa's natural environment present to those who would plan its development are discussed.

Chapter 4 examines the ways people use the resources of Africa's rivers and floodplain wetlands, focusing in particular on floodplain wetlands. It describes indigenous agricultural systems in detail, and the rich tradition of small-scale indigenous irrigation in Africa. This ranges from simple techniques that adapt cropping to natural flooding patterns through to practices that offer full water control. Figures are presented that show the surprising extent of informal small-scale irrigation in many African countries. Chapter 4 also discusses the importance of wetlands for fishing and grazing, the close integration between different kinds of economic activity and the way in which seasonally-flooded wetlands can sustain economic activities over large areas of surrounding drylands.

Chapter 5 discusses river basin planning. It explores in detail what has happened when developers have tried to import ideas about water resource development and implement them in Africa. It briefly describes the history of control on the Nile, the origins of the idea of river basin planning in the USA and its application in Africa. The work of the river basin development authorities in Nigeria is analysed in detail, and some of the very significant problems associated with their projects are described. This experience is then used to explore the factors that underlie the continued enthusiasm for river basin planning. The development of dams and irrigation schemes have been remarkably unsuccessful, and yet they continue to be built. The chapter explains why, and outlines the powerful factors that continue to promote their development.

Chapter 6 then turns to the problems of dam construction in Africa. It describes the problem of population resettlement, and the impacts of dams on floods downstream. Changes to flooding patterns have significant impacts on natural ecosystems of downstream wetlands, on agriculture and fishing. The chapter describes the growing conflicts over resource use in African floodplains, and the ways in which these can be exacerbated by development. It then looks at the technical methods used to appraise projects, cost/benefit analysis and environmental impact assessment, and asks why they have not been effective in ending adverse environmental impacts.

Chapter 7 discusses irrigation. It presents statistics on the extent of irrigation in different African countries, and discusses the experience of the large-scale irrigation projects built in a number of countries through the 1970s. These projects were extremely unsuccessful, expensive to build and to run, and unpopular with participants. The chapter therefore breaks down the problems of project planning and management to look in detail at the reasons for their failure. It then looks at small-scale irrigation, which is widely seen as a sound alternative. It argues that in fact many small-scale irrigation schemes are just as unsuccessful as larger projects, and it explains why this is so. Despite the poor record of many irrigation schemes, development agencies and African governments continue to see an important place for irrigation in development plans. The chapter ends by discussing implications for the rural poor of one possible future directon for irrigation policy, the development of large-scale private irrigation projects.

The final chapter explores alternatives to large-scale centrally-planned development. It discusses the possibility of developing indigenous irrigation systems, and the prospects for the rehabilitation of existing large scale irrigation schemes. Conventional approaches to development are not effective in either case. The chapter then turns back to wider questions about river control, and explores an alternative to large-scale flood control by dams, the idea of releasing predictable controlled floods to meet the needs of people downstream. It describes how this might work, and the progress of experiments with the technique. Finally it discusses the insights of the International Union for the Conservation of Nature and Natural Resources (IUCN) which is working on the sustainable use of wetlands in a number of countries.

The book ends with a discussion of the need for new approaches to development. Large-scale bureaucracies looking for standardized ways of approaching development 'problems' are likely to repeat all the mistakes of the past. What is needed is not just a new target for development planning, or new ideas, but a total break with the hierarchical top-down outsider-dominated planning that is still standard practice today.

Much of this book looks backwards, to review the performance of past projects. This is not because there is any particular satisfaction in recalling old failures or using the twenty-twenty vision of hindsight to criticise past planners. It is simply that the chief failure of outside development agencies and those who work for them has not been that they make mistakes, but that they have (by and large) been so bad at learning from them. Furthermore, many of the mistakes of the past were caused by attitudes, ideas, institutions and economic structures that are still current. Standard approaches to project planning have proved inapplicable and unsuccessful. The same approaches and institutional structures are unlikely to work much better in the future, no matter how their targets are re-defined.

The future for African rural people must be based on the informal skills of local people, organized and directed in concerned political and practical action by those people themselves. Development is what those people will do with the resources and ideas at their disposal. Development planning must be something they control. Outsiders must come as equals to meet with local people face to face, and must seek to facilitate and not to dominate. This is an idealist prescription, and it takes no visionary to see that it will not be easily realized. For all that, it may now be time to try.

# CHAPTER TWO

# Changing Africa

*Guthninga wa kuura*
*(Having clouds is not the same thing as having rain)*[1]

## Africa discovered

Only in recent centuries has European knowledge of tropical Africa extended beyond the coastal fringe, and interest expanded further than the supply of ivory, slaves and other items of trade. In Victorian times, Africa was the 'Dark Continent', into which heroes and hustlers penetrated in the name of commerce, faith or country. They found a remarkable world, diverse and difficult, and to it they brought vigorous agendas for change. After the explorer and missionary, soldier and merchant came the administrator of *pax europea* and his armoury of weapons of civilization: taxation, education, science and western justice. By and large the imperial newcomers failed to appreciate its cultures and political systems, the nature of its history or its environment. Africa was treated as a clean slate on which colonizing nations could write a version of European 'civilization'. In the view that A.G. Hopkins so neatly parodied, 'Africa's release from barbarism waited until the close of the nineteenth century, when the Europeans came, like cavalry over the hill, to confer the benefits of Western civilisation'.[2] Among historians, the elitism, indeed the incipient racism, of old imperial history has by and large long gone[3] (although something of the same myth endures among consultants who globetrot from project to project, coming out when they stop to drink beer and swap tales). There is now a better understanding of the nature and impacts of colonial rule. Something of the revolution in thinking is captured by the title of Walter Rodney's classic book '*How Europe Underdeveloped Africa*'.[4]

Certainly the colonial experience and the centuries of exploitative trade which preceded it are commonly seen to have had an overwhelming impact on the nature of African society, economy and (of increasing interest to African historians) environment. The classic study taking this line is Helge Kjekshus' book *Ecology Control and Economic Development in East African History*.[5] This argues that imperial occupation disrupted a prosperous and intricate economy in pre-colonial Tanzania, and also upset established patterns of environmental management, or 'ecology control', in particular of tsetse fly. This argument fails to take previous work on the ecology of tsetse fully into account, notably John Ford's already published book *The Role of Trypanosomiasis on African Ecology*.[6] Furthermore, John Iliffe argues that Kjekshus contributes to what A.G. Hopkins called the 'myth of Merrie Africa', that:

> the pre-colonial era was a Golden Age, in which generations of Africans enjoyed congenial lives in well-organised smoothly-functioning societies. The means of livelihood came easily to hand, for foodstuffs grew wild and in abundance.[7]

Africa has sometimes been taken to have been some kind of Garden of Eden,[8] in a state of harmony which the arrival of the imperial powers disrupted severely, reducing the indigenous peoples through 'ruthless exploitation' to 'a degree of poverty they had not known in the past'.[9] John McCracken, writing about Malawi, argues that such ideas oversimplify both the impact of colonialism and of the spreading capitalist economy.[10] A richer understanding of both pre-colonial economy and society and colonial impact are needed.[11]

Other studies confirm the complexity of the impacts of new political and economic structures on the people of rural Africa and their environments. The work of Michael Watts on the political economy of colonial northern Nigeria, for example, traces the impact of new economic relations as the area was locked into the world economy.[12] Colonial rule brought taxation and transportation, cash crops (cotton and groundnuts) to replace food crops and the erosion of the old moral economy of reciprocity and mutual responsibility between rich and poor. Falls in groundnut prices squeezed northern Nigerian farmers and exposed them to the impact of drought from the late 1960s onwards. Similar processes occurred elsewhere in West Africa.[13]

The impact of what Alfred Crosby calls the 'ecological imperialism' of European colonisation, was limited in Africa by environment and disease that restricted European crops and settlement.[14] However, not only the destruction brought about by centuries of slaving, but the colonial experience itself had a huge effect on African economy and society. Much of it was far from benign. Geoffrey Gorer, writing in the 1930s about a trip though French West Africa, described the coercive side of the 'civilising' influence of the colonial state: taxation and forced labour. He predicted gloomily that the African might go the way of the American Indian, a reserved minority on their own land.[15]

One of the most far-reaching impacts of contacts with peoples from outside Africa was more benign: the introduction of new crops from both the New World and Asia. Work in the Ivory Coast identifies a wide range of domesticated plants introduced before this century from the New World.[16] Many of these have become staples in Africa, integrated into 'indigenous' production systems. They include cassava, ground-nuts (peanuts), tomatoes, maize, sweet potato, cocoyam, papaya, to-bacco and peppers. From Asia came Asian rice, taro, bananas, sugar cane, mango and citrus fruit. The same story is true over most of Africa. The remarkable and diverse banana culture of Buganda, for example, is based on an introduced species, although its date and route of arrival remain obscure.[17] Much of what looked like an unchanging 'traditional' system of production when colonisers arrived late in the nineteenth century was of relatively recent origin.

This kind of work shows clearly the complex historical context of contemporary Africa and the vast range and scope of outside impacts. In particular it is revealing about the way relations between people and environment over time change, and about the impact on that relation-ship of the traditional stuff of history, relations between people. The rise of a specific environmental history has important implications for our understanding of Africa.[18] Neither African societies nor environ-ments have been static in the past. Development agencies are not the first or the only agents of change in rural Africa, and their projects are just the latest outside impositions on the environments and peoples of the continent.

## Economic development in Africa

Not much was left unchanged by this onslaught by the time African countries began to demand and win independence after the end of World War Two. In the 1960s new countries appeared on the map at a tremendous rate. Mostly they did so with a nominal democracy and some pretence at independent government and economy. Most of Africa has now been independent for some two decades. Indepen-dence has accelerated the cultural and economic transformations of the colonial period, and the forces which drove them. Yet it has not brought the prosperity that many hoped for. In the early 1960s, the French agronomist René Dumont wrote a fierce critique of develop-ment in Africa, *L'Afrique Noire est Mal Partie* (translated into English as *False Start in Africa*).[19] This targeted the record of both the colonial powers and the elites of newly independent Francophone countries.

He wrote: 'African development problems can be summed up in one word: underdevelopment'. After centuries of shameless exploitation, metropolitan countries had expressed a desire for development in Africa, but their aid programmes 'remained for a long time – and still are – badly conceived and administered'.

In the 1980s Africa was commonly portrayed as locked into a chronic trap of poverty, corruption, warfare and environmental degradation. The picture was a little different twenty years before, as African nations stepped into independence. Then the conventional wisdom was that Africa had a bright economic future. This view was shared by the new governments, determined to sprint ahead to catch up with the wealthy and industrialized North, and their advisers from old colonial nations and new multilateral agencies like the World Bank. Economic growth, improved infrastructure (such as roads, education, health services) and improved standards of living seemed believable goals. Colonial rule had, in most cases, moved economy and people only a very short distance down the road towards such goals (indeed, many argue that it did much to set such 'development' back, and in some ways this is probably true). However, nation-building and investment in economic infrastructure (mostly using foreign capital, the now-ubiquitous 'aid') proceeded apace. Over two hectic and at times desperate decades later, the results are far from impressive.[20]

At first glance, economic growth has been remarkable, on average 3.4 per cent per year since 1961.[21] However, growth has been uneven, both over time (with real economic decline in the 1980s, for example), and between countries. Although there was significant industrial growth in the 1960s, the 1980s saw de-industrialization begin. Africa's share of export markets has fallen by 50 per cent since 1970, and its share of world trade fell from 2.4 per cent in 1970 to 1.7 per cent in 1985. If the value of oil exports from the few African oil-producers is taken into account, the figure would be even less. Africa is still hugely dependent on the export of primary commodities, including oil. These represented 88 per cent by value of exports in 1988.

For many African countries serious economic and financial problems began in the mid-1970s. Like other oil-exporting countries, Nigeria did well with high oil prices in the 1970s, but its economy slumped in the 1980s as prices fell. It borrowed heavily against future oil revenues, and now has a major problem of debt (see Figure 2.1). Total external debt in Nigeria rose from $8.8 billion in 1980 to $29.8 billion in 1987. This was 10 per cent of the annual value of goods and services exported. In oil-importing counties like Kenya, high oil prices and declining value of raw materials and commodities on the world market has made conditions difficult over a long period. Kenya's debt was far less than Nigeria's, at $5.9 billion in 1987, but this was 29 per cent of the value of exports. Several African countries, such as Guinea-Bissau, Madagascar, Burundi and Niger suffered debt burdens more than 30 per cent of the value of exports. Although the total indebtedness of some of the richer countries (like Nigeria or Ivory Coast) is very great, it is the poorest countries which face the harshest problems in terms of repayment. Many African countries failed to service their debts through the 1980s.

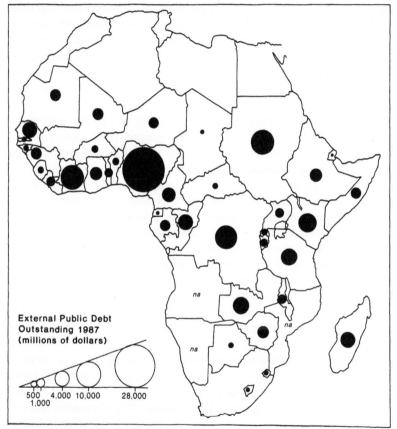

External Public Debt Outstanding 1987 (millions of dollars)

500   4,000  10,000    28,000
   1,000

*Source:* World Bank (1990) *Sub-Saharan Africa: from Crisis to Sustainable Growth*, Washington

**Figure 2.1**  Debt in sub-Saharan Africa

In retrospect, it is clear that the newly independent African countries were unwise, and badly advised. They pushed for 'modernization', uncritically copying Northern models. Development was seen to be something driven from the top, and was dominated by a drive for industrialization. Agriculture was squeezed, receiving little investment while being taxed to finance industrial growth. Where agriculture was targeted (when agricultural export declines and food production shortfalls began to bite), peasant farmers were bypassed in favour of grandiose mechanized development schemes. Industries grew, but more slowly than urban populations as rural producers sought better opportunities. Governments lacked the technical and managerial skills to make a success of development projects, to choose between the opportunities offered by aid donors and foreign companies. Between incompetence and outright corruption, many industrial development projects and infrastructural works (roads, dams, harbours) failed to work properly. Others worked in a technical sense, but were left unintegrated into the wider economy. They left a legacy of massive loans to be repaid. The informal sector (and the black economy) flourished, and peasant farmers retreated from the clutch of market economy and state bureaucracy.

According to the World Bank, the result of all this has been weak agricultural growth, declining industrial output, poor export performance, growing debt, declining institutions, and deteriorating socio-economic and environmental conditions. Infrastructure which has been put in place over the last 30 years – at enormous cost – is deteriorating. Maintenance has lower priority than new projects, both among African politicians and aid donors. It looks better to build a new road than to repair an old one. Roads, railways and universities are in decline, just when they are most needed. Furthermore, much of Africa has been ravaged by war over the last twenty years. Zaire suffered a crippling civil war after independence, as have Ethiopia and Somalia in recent years. Destabilization and warfare have cost Angola and Mozambique deeply in both human and economic terms. Large areas have been devastated, and millions are homeless. The human and economic costs of conflict and destabilization are vast. Quantifiable costs in southern Africa are estimated to amount to 25–40 per cent of GDP.[22] In Mozambique, 3.3 million people (23 per cent of the population) face severe food shortages, 1.1 million have been forced to leave their homes and 0.7 m have fled the country.[23] It is a similar story in Ethiopia and the Horn of Africa, where civil war is the most insidious and terrible source of famine. Instability and warfare are widespread in Africa: Somalia, Sudan, Chad, Liberia, Uganda, many countries are affected. Even in countries nominally at peace, military spending exerts a crippling burden on government expenditure.

Furthermore, economic growth in Africa has been overtaken by population growth. Life expectancy at birth has improved, and is likely to continue to do so. Crude death rates have fallen from 23 per thousand in 1965 to 16 per thousand in 1987, but birth rates have barely fallen at all (48 per thousand in 1965, 47 per thousand in 1987). Levels of infant and child mortality have fallen, as has maternal death in childbirth, although both remain staggeringly high by Northern standards. Average life-expectancy in Africa in 1987 was 51 years. Over half the population of the Sahel is under 15.

Family planning programmes have proved expensive (both in terms of money and skills), and slow to have any effect on birth rates. However, family planning has the potential to allow women to space and time births, with direct potential benefits in terms of mother and child health. More African governments are starting to take family planning seriously. Francophone West African countries, for example, have recently been repealing French colonial laws banning contraception. Chad did so only in 1988. Total fertility rates are highest in some of the poorest countries, particularly in the Sahel. In 1950 the total population of the nine countries of the Sahel (Somalia, Ethiopia, Sudan, Chad, Niger, Mali, Burkina Faso, Mauritania and Senegal) was about 46.9 million. By 1980 it had almost doubled to 91.5 million. The United Nations predict that it will triple again by the year 2020, reaching 263.3 million. The current rate of growth is 2.8 per cent, but this is rising, and will do so for at least the next decade.

Overall population densities remain low in sub-Saharan Africa (compared to South or East Asia for example), but many areas are inherently unproductive, and others are shattered by conflict. The key problem in terms of development strategy is the *rate* of population growth. In countries such as Kenya, the rate of population growth has been above 4 per cent per year. There are now 450 million people in Africa, twice the number at independence. By 2010, there could be one billion. The provision of basic services (shelter, food, health, education) is already expensive for these poorest countries, racked by debt and falling exports. Meeting the new needs of a burgeoning population will be very difficult. As Anthony O'Connor comments, 'to provide a substantially improved standard of living for so many would be little short of miraculous'.[24]

## Agriculture in sub-Saharan Africa

Such miracles are not in immediate prospect. Indeed, in many ways Africa is running just to stand still. Agricultural output has grown, but on average by only 1.5 –2 per cent per year. The value of agricultural exports has declined. The rate of growth in food production has been less than this, and far less than population growth. Food production per capita fell in the 1970s, and despite recovery in many places it remains depressed (see Figure 2.2). The 1980s saw many countries in Africa in serious food deficit. United Nations data suggest that six of the nine Sahel countries produced less food per capita in the early 1980s than in the period 1974–76 (just after the Sahel drought). Drought and war are creating widespread famine, and food imports have been rising by 7 per cent per year. In 1986, cereal imports in Africa amounted to 8 billion tonnes. Despite this, malnutrition is widespread.

## Index of food production per capita
### 1961–65 average = 100

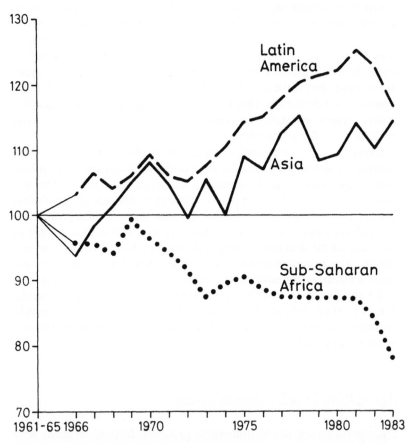

*Source:* World Bank (1984) *World Development Report,* Oxford University Press.

**Figure 2.2**  Food production per capita in sub-Saharan Africa
Most commentators look to urbanization and industrialization to absorb these extra workers and consumers. The infrastructure needs of

this transition are vast. Urban growth has been rapid in Africa in recent years, and is rising by almost 7 per cent annually. In some countries significant proportions of the population are already urban people, for example Zaire (38 per cent), Gambia (36 per cent), Mauritania (38 per cent) and Nigeria (33 per cent). By 2020 there are likely to be 30 African cities with more than one million people. Urban areas in Africa are already swamped in terms of the provision of basic needs, particularly in the poorest countries such as those of the Sahel. Rural-urban migration, and migration between ecological zones (and between countries) is expected to increase. The political and economic implications of this are likely to be dramatic.

However, despite the rapid urbanization of recent years, the vast majority of Africa's people are still rural. The rural environment provides them with agricultural crops, harvested resources (such as fish) or grazing for their livestock. The wider economic fabric of rural Africa depends on the capacity of the environment to sustain these productive activities, and through them the craft specialists, artists, traders and others who in their turn depend upon primary producers. World Bank figures suggest that agriculture supports 66 per cent of Africa's workforce, and produces 33 per cent of Africa's GDP and 40 per cent of exports by value.

Maintenance of the productivity of the rural environment is vital to Africa's economic future for two reasons. First, many national economies depend on the export of agricultural commodities. There was much debate about the World Bank's report in 1981[25] which argued that Africa should continue to specialize in primary products, and particularly agricultural commodities like tea, cocoa, coffee, edible oils and cotton, but there was little doubt about the dependence of economies on these products. Second, productive and attractive rural economies are vital if the rate of rural-urban migration is to be kept within limits which infrastructure, employment and service provision might be able to handle. Such migration creates many problems in both receiving and supplying areas: most such migration is by economically active young men. The communities left behind are dominated by old people, women and children. Such communities are likely to lack the labour and capital necessary to maintain existing agricultural systems in good heart, let alone to innovate and adopt new ideas and technologies.

For many people in Africa, hope for the near future at least rests on rural production. Migrants to urban areas may encounter human and environmental degradation of an order unimagined in their home area, for the streets of African cities are not paved with gold. Furthermore, for an increasing number of rural households urban incomes are as important as rural resources in achieving a secure livelihood. Thus for pastoralists from the Red Sea Hills in Sudan, wage earning in the docks of Port Sudan (the main port for off-loading famine relief supplies for Sudan) has been vital to the maintenance of the nomadic livestock economy.[26] Many households practise a double-strategy, with members in cities and on the land supporting each other through economic and environmental crises. Rural production can have a vital role in supporting urban workers, particularly where infrastructure and economies start to collapse, as they did in Ghana and Uganda in the 1970s. Rural Africa has a vital role to play in the economic future of the continent.

There has been a great deal of research within the Third World over the last 25 years aimed at improving the productivity of agriculture, at achieving a 'Green Revolution'. The key element within this has been the development through plant breeding of a series of new varieties of crop plants that are more responsive to fertilizers and give higher yields. This work has been done primarily within the public sector in the International Agricultural Research Centres (IARCs), for example IITA at Ibadan in Nigeria, IRRI at Los Baños in the Philippines and ICRISAT in Pakistan. Through these centres, higher-yielding and more fertilizer-responsive varieties of several major crops (particularly rice, wheat and maize) have transformed food production prospects in some Third World countries, notably those of East and South Asia. There has been criticism of the impact of this package, for example that it helps rich farmers more than poor farmers or landless labourers, but its impact has certainly been significant. However, the Green Revolution has to a large extent passed Africa by.

There are numerous reasons for this.[27] The crops which have received most attention from plant breeders, wheat and rice, are not mainstream food crops in Africa. They comprised 80 per cent by weight of cereals harvested in South and East Asia in 1982, but only 13 per cent in Africa. Development of the main African food crops (millets and sorghums, yams and cassava) has been much slower, and less successful. Research on Africa has been hampered by low government expenditure, just as agricultural output has been held back by poor infrastructure and unhelpful government policies. It is not possible simply to transfer experience with (for example) improved varieties of sorghum and millet from South Asia to Africa, where climatic, soil and institutional conditions are quite different. Furthermore, in comparison with Asia, relatively little is still known about the complex production systems of Africa, and data on what crops are grown where is very poor. Without such data, plant breeders cannot design sensible research programmes. Research programmes driven by researchers', perceptions of farmers' problems and carried out in research stations and experimental farms are likely to be remote from the real-world needs and potentials of rural Africa.[28] Such research is unlikely to lead to dramatic leaps in the productivity of African food crop production in the immediate future.

## Experts in Africa

The development of Africa has always been driven by outsiders. This is true not simply in the sense that economic power in colonial and independent Africa have lain outside the continent, but also in that power to define goals, to make plans and spend money, indeed power to define the meaning of development itself, was in the hands of strangers. Rural Africans without formal education lie furthest away from the centre of such power.

Even today's African government decision-makers are outsiders to the remote rural environment, for they are remote from it and often quite ignorant of it. African politicians and bureaucrats rarely visit rural areas, and when they do they are insulated from them in motorcades and air-conditioners, and limited to what can be seen from aircraft, helicopters or metalled roads. Politicians and government bureaucrats build their own images of environment and poverty in rural Africa from past experience (perhaps in rural areas, perhaps in First World Universities) and the images of rural areas prevailing in capital cities. In this they have much in common with aid agency decision-makers. For details of ecological and socio-economic trends, both depend on the reports of consultants. These 'experts' usually come from the First World. They have their own preconceptions, built up from their education, the views commonly accepted by their peers, and the ideas which have proved acceptable to other clients in previous missions. Like the bureaucrats they serve, they have little chance to challenge these assumptions.

Fieldwork for consultants is usually brief, constrained by timing and access. The inherent biases in their work are described by Robert Chambers. They include, for example, seasonal biases (fieldwork is very difficult in the wet season), 'tarmac' biases (few venture far from main roads), and biases on a few over-studied accessible projects.[29] Their task is to get in and get out as rapidly as possible, getting just enough data to be able to make (and defend) policy recommendations. Their professionalism tends to be at war with their business sense. Of course, consultants' recommendations have to be sound enough that they are re-employed. It is often (rightly) said that consultants are only as good as their last job. However, there are few consultants who do not feel constrained by time and terms of reference. Of course, some consultants have a remarkable fund of knowledge about rural Africa, and may have the courage to buck the received wisdom about environment and development where they believe this to be necessary. Sadly, but unsurprisingly, some do not. Their voluminous reports serve to strengthen the status-quo and fashions in thinking about development in rural Africa. Where that thinking is wrong-headed, consultancy reports entrench misconceptions.

There is a growing literature on the importance of experts in shaping Africa.[30] The word 'expert' is used extensively in this book, sometimes in the ironic sense of someone who indeed presumes to know all the answers (and is allowed great power to plan on that basis) but who makes mistakes.[31] Experts wield great power to transform the lives of other people. Despite professional skills and good intentions, this power does not always work for the universal good. Adrian Adams, writing of the Senegal Valley, says:

> it is not knowledge or skill alone that's wanted of an expert; there would be less costly ways of acquiring them. What matters is the halo of impartial prestige his skills leave him, allowing him to neutralize conflict-laden encounters – between governments, between a government and its governed – and disguise political issues, for a time, as technical ones. An expert helps disguise the government of men as the administration of things, thus making it possible for men to be governed as if they were things.[32]

If this indictment is even partly true it suggests a need for honest self-appraisal. It is important to be aware of both the power and the limitations of outsiders and their expertise.

In the past, engineers, economists, agriculturalists and administrators have believed that their image of Africa was accurate enough that they could prescribe solutions. The first 'experts' were colonial administrators, followed from about the 1940s by a cadre of technical officers. After independence, their place was taken by international consultants and the employees of aid donors. There is now a vigorous international development community of engineers, hydrologists, economists, agriculturalists, and sociologists. They move from project to project, implementing dreams and making plans; disbursing money and building structures in the bush. They answer to politicians in First World countries (who control the purse-strings) and to decision-makers in African governments, many of whom lack training in development, or who received it before the failures of the 'sprint for modernization' became widely known.

## Indigenous Technical Knowledge

There is much written about Indigenous Technical Knowledge (ITK) in Africa.[33] Some of it is uncritical, using idealist views of 'traditional' agriculture as a stick with which to beat modernizing scientific approaches to the African environment. This is a lively critique, although it is not always soundly based. Other studies have played a vital role in reclaiming into the development debate some understanding of the skill, flexibility and dynamism of African agriculture, and the possibility of 'indigenous agricultural revolution'.[34] African land users have often had (and still have) a better understanding than anyone else of what can and cannot be done in their environment. This understanding stems in part from their grip on the past. They are not tied to scientific knowledge, or to the history in books. Developers from outside do not have the ears to hear this experience of the past, and that is why development projects so often ignore local knowledge of environment. It is this which the phrase 'ITK' attempts to capture, and bring to the attention of the outside expert.

In the context used in this book, indigenous knowledge embraces specific techniques of the management of land and water resources (flood cropping or hill furrow irrigation, for example; see Chapter 4) and also the cultural and socio-economic systems in which they are rooted. It is thus used as a form of shorthand for a whole range of things which people do and ways in which they think. It embraces existing technology, technical creativity and the capacity to organize and be creative in response to changing circumstances.[35] But what does the word 'indigenous' imply? Certainly not that such knowledge is ancient or unchanging. John Sutton writes 'there persists the notion of an essentially unchanging, a supposedly "traditional" African past, the details of which need little researching, since they can be readily assumed'.[36] That is an unhistorical, and mistaken, view, and is one reason for avoiding the word 'traditional' which seems to imply archaic knowledge of a kind that is by definition not appropriate to contemporary conditions. Modernizers have persistently characterized such 'traditional' ideas as a barrier to 'progress' and 'development', and the word carries many negative connotations.

Indigenous systems are not to be seen as primitive in some way, or as uninfluenced by ideas and events either at the present time or in the recent past. Rather, the word 'indigenous' implies knowledge and practices which are locally acquired and controlled, rather than imposed from outside. Of course, it carries the connotation that such ideas are somehow more 'rooted' in local culture and economy, but this does not mean that they cannot stem from remote first-world science, or even the programmes of development agencies. What distinguishes indigenous knowledge is not its source, or its age, but the form of control upon it and the extent to which it is appropriated, disseminated and applied by local people. Indigenous knowledge might incorporate ideas and practices of great age, but these could be mixed with ideas proffered by colonial agricultural officers or World Bank, consultants or (increasingly commonly) from other farming or pastoral groups contacted through international non-governmental organization networks.

What, then, can this knowledge teach us? First, that existing land use systems are flexible and fluid through time. Robert Chambers writes that 'complexity and diversity are dimensions of the livelihood strategies of the rural poor'.[37] Second, they are highly adaptive, both to environmental change and to changes in society and economy. Third, that this adaptability leads to great resilience. Resilience is an ecological concept, a characteristic of semi-arid ecosystems among others. It is also a feature of cultures and economies which have been developed in such ecosystems. Adaptability and resilience (culturally and economically) are, of course, characteristics which have helped African producers to survive the many impacts of change created by an expanding world economy and mediated by colonial and post-colonial states. The resilience of African farmers and pastoralists, and the characteristics of the difficult environments that they inhabit, are the subject of the next chapter.

# CHAPTER THREE

# The hard earth

*The teacher of a culture is its environment, and agriculture is its classroom*
*P. de Schlippe (1956)* [1]

## Using Africa's Drylands

One of my first impressions of Africa was the view from the window of
a Nigerian Airways Fokker jet as it cruised west from Kano towards the
city of Sokoto. Beneath us, the face of Africa was as human-made a
landscape as the suburban streets of London and the green fields of
Sussex I had left the day before. The ground was etched out in a dense
network of fields, their boundaries marked by lines of low bushes, and
their surfaces punctured by spreading trees. Every few miles was the
rounded shape of a compact village, its rectangular streets marked by
trees again, separating a complex hive of compounds. It lay like a
densely-packed island of humanity in an open sea of fields.

Later, I learned that the land around both Kano and Sokoto is
among the most densely populated in Africa. These cities are sur-
rounded by what is called a close-settled zone: permanently settled land
which is farmed year after year. Hausa people have farmed this area
more or less intensively for many generations. The descriptions of
Kano by the explorer Heinrich Barth in the nineteenth century are not
wildly different to what could be seen today, away from the main road
corridors and their advertising hoardings and roadside vendors.

Hausaland is not a particularly favoured environment for agriculture. Rainfall varies, but at 1,000–1,800 mm per year it is well above that
of the Sahel. Nonetheless, it all falls in a single rainy season between
May/June and September/October. Soils are infertile. In the close-
settled zones, sewage was packed out from the city on donkeys to be
spread on the fields along with animal manure from house compounds
and from the cattle of nomadic Fulani which pass through in the dry
season. Visiting the area again in December 1990, mounds of com-
pound sweepings were a common sight in the fields.

Walking those fields in my first wet season in Hausaland was quite an education. Like many outsiders before me, I was amazed at the diversity of crops grown, their shape (millet and sorghum crops with stems two or three metres high), and the way in which several crops were grown together in the field. Crops looked diseased, but did not die. They dried out, but at the end of the season there was some kind of yield. Farmers used the dryland, but where they could they also used wetlands along rivers. Farmers all seemed to have livestock, other jobs and craft skills. In fact, my very notion of 'farmers' had to be drastically revised as the importance of off-farm incomes and the role of women farmers slowly emerged.

At first the whole thing looked like chaos, very foreign to an eye trained on the groomed monocultures of eastern England. Gradually the logic of it sank in as I read more and talked to farmers about what they planted, and when and why. By the end of a couple of years I was starting to understand the physical and economic environment against which these farmers pitted their skills and knowledge, and in particular their responses to drought.

## Uncertain rain

Much of the debate about water and development in Africa turns on an assessment of patterns of rainfall.[2] The fundamental question is simple: are the droughts of the last 20 years 'natural' or caused by human use? This demands that we know a great deal more about the way in which rainfall is distributed in space and time.

Much has been written on rainfall in Africa.[3] The total amount of rainfall and its seasonal distribution go a long way to explaining patterns of hydrology, vegetation and human land use. Variations in that precipitation are a major threat to food security and life.[4] Broadly, air moves towards the equator across Africa, from the southeast in the southern hemisphere and from the northeast in the north. Around the equator this air warms, and tends to rise, and return at high altitude in the atmosphere towards the poles. Where this air descends again (for example over the Sahara) it is dry.

The highest rainfall in Africa occurs on the western side, particularly in Liberia, southern Nigeria and Cameroon (see Figure 3.1). Mount Cameroon can experience 10 metres of rainfall annually. There is also high rainfall through much of the coastal zone of West Africa, the Zaire basin, the mountains of East Africa and Madagascar. Within the general pattern, rainfall varies markedly with altitude. There are two arid areas, the Sahara and the Namib deserts. Rainfall increases southwards from the Sahara through the Sahel (250–600mm) and zones of increasingly wet and wooded savanna (the Sudan and Guinea Savanna Zones) towards the tropical forest near the coast. In south-central Africa, rainfall increases towards the northeast away from the Namib. The vegetation here is mostly savanna woodland.

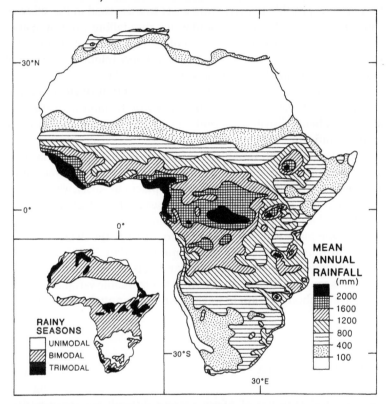

*Source:* Nicholson, S.E., Kim, J. and Hoopingardner, J. (1988) *Atlas of African Rainfall and its Interannual Variability*, Florida State Univesity

**Figure 3.1** Mean annual rainfall in sub-Saharan Africa

Most of Africa's rainfall comes from moist air brought in from the Atlantic. Westerly winds provide rainfall over the vast Zaire Basin in the northern hemisphere winter. These moist westerly winds move northwards towards the Tropic of Cancer through the year until they reach their furthest northwards extent in the northern Sahel in July. These areas near the tropics receive a single rainy season during the northern hemisphere Summer. Areas nearer the Equator (for example East Africa) have a dual rainy season: often called (rather inaccurately) the long and short rains. Patterns of rainfall in east Africa are somewhat complicated by the effect of the Indian Ocean trade winds.

Rainfall in Africa is not only seasonal, but is also highly variable in space and time within one season. Much of it falls in the form of short, intense, convective storms. These are rather like summer thunderstorms in Europe. The edge of these storms can be very well-marked. It is perfectly possible to drive from a wet to a dry road in the space of a few yards, or stand in the dry and watch rain falling close by. There are many implications of this pattern of rainfall. It means that rainfall is very unpredictable on a day-to-day basis. There might be 100 mm of rainfall in any particular month at a site in the Sahel, for example, but all of that may fall on three or four days in a series of storm events. Furthermore, it may all fall at the beginning of the month, or the end, rather than being distributed evenly. The implications of this for farmers are significant.

Rainfall is also highly variable over short distances, such that farmers in one area may receive much less rainfall in a given week than those just a few kilometres away. There are few places in Africa where there are enough rain gauges to be able to measure this effect, although it must be well known to anyone who has stood and watched rain come and go. One place where it is possible to measure the detailed spatial pattern in rainfall is the Gezira Scheme in Sudan. This is a massive surface irrigation project on the flat clay plains between the White and Blue Niles south of Khartoum. There is a remarkable variability across a distance of about 100 km in weekly and even annual rainfall totals.[5]

## Dryland agriculture

The semi-arid savannas of Africa present difficult environments for agriculture. The success of crops depends on the length and adequacy of the rainy season. In humid areas with only a short dry season, rainfed agriculture is relatively unproblematic (except for problems such as soil erosion and fertility), and a diverse range of crops can be grown. As annual rainfall declines (e.g. northwards towards the Sahel), the range of crops that can be grown declines, as do yields per unit area. Farmers plant larger areas with crops spaced further apart to maintain yield per unit of labour as productivity falls. Areas with an annual rainfall equivalent to that at the northern edge of the Sahel (250 mm) are marginal for agriculture. Here, relatively minor fluctuations in rainfall amount and distribution have a very significant impact on crop yield. Research in the 1980s has shown that it is not just the length of the rainy season that is important, but also the distribution of rainfall within it. Breaks in the rainy season, or periods with little or no rain, can have disastrous effects on crop yield, even if the total amount of rainfall changes very little.

Variability in the timing of the start of rainy seasons, or the occurrence of dry periods within them, represent major sources of uncertainty for farmers. If they plant too early, in response to an early ephemeral storm, the crop (and the seedgrain) may be lost. If they wait, and miss part of the period of rains, they risk giving their crops a shorter overall growing season and miss the flush of nitrogen that occurs with the onset of the rains. Statistical studies of the characteristics of wet seasons in the central Sudan through the past century show that they are shorter in dry years. In the dry period of the 1960s and 1970s there was a particular fall in the reliability of mid-season rainfall. This had serious implications for farmers, since crop losses at that stage could not be circumvented by replanting.[6]

Rainfall is by no means the only factor in the equation. Plants lose water through their leaves in transpiration. Only in the wet season does the amount of rainfall exceed the evaporation losses from both open water surfaces and plant leaves. Without sufficient water, plants wilt and rapidly die. Soil conditions (texture, depth and water infiltration capacity) can have a major impact on the amount of water actually available to crop plants. Dry periods within the rainy season can cause wilting and loss of yield. Wild plants in semi-arid regions often have adaptations to prevent water stress in the dry season, and extensive root systems to acquire water from a greater volume of wetted soil.

High temperatures mean high rates of energy loss through respiration. Tropical plants have a lower rate of growth per unit of solar energy they receive than plants in temperate areas. Daylength varies very little near the equator, so in the growing season days are fairly short compared to temperate areas, and temperatures (and hence respiration rates) remain high at night. A number of tropical plants have a different metabolic pathway for photosynthesis which gives them an advantage in such conditions. Crops such as sugar cane have this characteristic, but it only seems to favour plants which grow where there is relatively plentiful water.

The high temperatures in Africa are themselves a constraint on the adoption of crops from temperate regions. Wheat, for example, will only grow in the savannas of Nigeria in the dry season when temperatures are lowest. The date of planting is vital. There is a short 'window' of time during which wheat can be planted with the expectation of reasonable yields. Outside that, it is too hot during the flowering period, and yields fall rapidly.[7] Although wheat (under irrigation) is being grown extensively in several parts of tropical Africa (notably with Canadian government aid in Tanzania), it is not suited to all parts of the continent. This is (at least in the eyes of developers) unfortunate, since it is temperate crops (grain crops like wheat or vegetables like the potato for example) which have been most extensively bred by scientists to produce high-yielding varieties. Of course, some improved varieties of other crops have been widely adopted, for example hybrid maize in Kenya and Zimbabwe, and there is now increasing research to develop improved varieties of sorghum and millet for semi-arid eastern and southern Africa.[8]

Although many of the crops grown in semi-arid regions of Africa are relatively recent arrivals, there is good evidence that some species were domesticated within Africa. In West Africa these include African rice (apparently domesticated in the Niger Inland Delta and still grown quite widely in West Africa), millet and sorghum. Ethiopia has a range of wheats, the grain crop *teff* and the African banana *ensete*. A number of these cultivars have ecological characteristics which make them well-adapted to aridity. Thus millet and sorghum both have waxy leaves which cut down water loss, and are also able to withstand long periods under water stress.

Cropped varieties have also been selected for short growing seasons. There are 'finger' millets grown in the Sahel which mature in 80–90 days. Such varieties have a far greater chance of reaching maturity within a short dry season, although short season varieties are vulnerable to short-term stress at critical points in their maturation. A farmer depending entirely on short-season varieties might lose everything in a short dry spell that is badly timed. Many indigenous varieties are more drought-resistant than alternative crops like maize which have often been proffered as part of development project packages, or higher-yielding varieties of the same species. Attributes such as drought resistance are only slowly being understood and translated into effective research aims. Crop breeding to select for short growing seasons has its drawbacks.

However, the adaptation of African agriculture extends beyond choice of crop. The use of mixtures of crops is common, both in the form of intercropping (simply growing crops together) and relay cropping (growing two crops together which have different harvesting dates, so that when one is harvested the next takes over) . In northern Nigerian Hausaland, for example, it is common to find millet, sorghum and a third crop planted together. This third crop might be cotton, groundnuts or cowpeas. This system has numerous advantages. First, it thickens the plant canopy and extends the period during which the soil is protected from potential rain-splash erosion. Second, it allows crops of different ecological requirements to be grown together, so that the farmer has some insurance against a bad season. Thus in a good rainy season the farmer may get a good crop of both grains and cash crops. In a poor season only the most resilient millet might give a good yield.

The third advantage of crop mixtures is that they give a farmer a variety of crops. Different crops have different uses: millet and sorghum, for example, are both grain crops, but quite different foods are made from them. Millet is mixed with milk to make *fura*, while sorghum is used to make a thicker grain dish called *tuwo*. Pulses and groundnuts have a high nutritional value. Millet is particularly useful to Hausa farmers as an item of trade with Fulani herders. Both millet and sorghum stems are important sources of fencing and roofing materials.

The variety of crops is also useful economically. Grain crops can be eaten or sold, as can cowpeas or groundnuts, while cotton provides a cash crop. In addition to these crops, Hausa farmers also grow root and tuber crops, leaf crops for making sauces, cucurbits (melons and calabashes) and many vegetables. Furthermore, the landscape of Hausaland has been described as 'farmed parkland' because of the number of trees growing in the fields. Some of these are fruit trees (e.g. mango), others provide other foods, such as the leaves of the baobab which are a common ingredient in soup.

Crop choice involves much more than simply playing games with nature from a static portfolio of crops. Farmers will experiment with new crops (high-yielding varieties of wheat, for example, or Irish potatoes), and generally have a good idea of what will grow and where. They typically have a clear idea of the capacity of different kinds of soil to support different crops, and for example exploit small patches of low-lying or damp land with great skill. Hausa farmers also adapt their cropping patterns during prolonged periods of drought. Crops like cassava, for example, are widely known as 'famine foods', and are often planted in drought periods. Innovation is a key to survival. During the long drought of the 1970s and 1980s, Kanuri farmers in northeast Nigeria began to use a small melon called *guna* which they obtained from Niger.[9] This is high in protein and grew well in minimal rainfall. In 1990, this crop could be seen in many fields.

Indigenous knowledge about crops available and their ecological requirements represents just a few of the elements in the Hausa farmer's range of strategies for living. Most farmers combine agriculture with other activities. Many will have livestock, perhaps sheep or goats, perhaps cattle. Most men will also have some form of off-farm income. This might be a craft skill such as building or butchering, fishing or trade. Trading can be local or long distance, with some Hausa travelling far down to the south of Nigeria in the dry season. Other men will migrate in the dry season, again often to southern Nigerian cities, in search of work. They might end up as building labourers, or working on a farm in the south. There are also other dry season activities for men which are very important, such as Koranic teaching or praise-singing. It is worth pointing out that although women farm in Hausaland, they do not have access to the same range of activities. Households without adult males are often poor. Across Africa as a whole, of course, about 80 per cent of farm work is done by women.

In other respects the Hausa are not so untypical of Africa. The important point about all this is that it is highly misleading to assume that 'farmers' do nothing but farm. Just as there are quite intricate skills of farm husbandry, there is a complex social and economic framework within which farming fits. This enables farmers to survive, and often thrive, despite the uncertainty of the environment. In times of drought, activities such as dry season migration are intensified, and farmers also make use of 'famine foods' gathered from their local area.

Agriculture in Africa's drylands is risky, by the very nature of the environment. Farmers have many strategies to minimize this risk, or minimise their exposure to it. Above all, however, dryland farmers in many parts of Africa seek to spread their operations across different kinds of environments. Just as they seek to have a foot in more than one economic activity, they spread risk by exploiting more than one ecological zone. The most important environments open to them are wetlands. In Africa, wetland areas have been the focus for some of the most intensive and most productive agricultural activities. They are not only important for agriculture, but are places where many different kinds of economic activity come together. Agriculture, fishing and grazing are often most closely integrated in wetlands. This integration is described in the next chapter. Wetlands have also been the focus for many of the schemes to transform production in rural Africa through development. The nature of these proposals will be discussed in Chapter 5.

## Dryland pastoralism

Many areas of Africa are too dry to support rainfed agriculture, and here nomadic or semi-nomadic pastoralism has long been the dominant form of land use. Recent decades have brought about a revolution in the ways in which outsiders view African pastoralists and their strategies for survival in arid and drought-prone environments. Researchers and (more slowly) policy-makers have now begun to appreciate the logic of nomadic or semi-nomadic pastoralist practices. Indeed it is ironic that while indigenous pastoral groups are remarkably well-adapted to Africa's natural environment they are vulnerable to competition from outside interests, particularly development projects that take over areas of dry season grazing.

Rainfall seasonality has a significant impact on the way African pastoral groups make use of the environment. Remote sensing from satellites confirms what pastoralists have long known, that the movement of rainfall with the onset of the rainy season has a dramatic effect on plant production. The duration and length of the rainy season (or seasons) is a critical factor in the ecology of natural ecosystems and the nature and timing of human use of land and water resources. The ecological patterns driven by these continental-scale variations in rainfall are vast. Huge numbers of birds, for example, migrate south into Africa from the Palaearctic region (Europe and Eurasia) to winter. On a smaller spatial scale, the massive wildebeest migrations of the Serengeti ecosystem in East Africa are triggered by the availability of grass, itself a function of rainfall. As the rain moves north, so too does the location of available grazing, and so too do the wildebeest.[10] Where such migrations are prevented, as they are by anti-disease livestock fences in Botswana, large numbers of animals die.

The livestock herds on which pastoral communities subsist respond in very much the same way to the ecological changes brought about by seasonal rainfall. In the West African Sahel the Fulani move north with the rains in June and July, moving south again as the dry season approaches in October. The northward penetration of the rains in June, and their subsequent retreat in September determine the limits of the pasture growing season.[11] The seasonal pastures of the northern Sahel are highly productive, are nutritious and easily digested. However, they last for but a short period. In order to exploit them, pastoral groups need to be mobile. Such a strategy allows them to time and direct their movements to make best use of available resources in any particular year.

Mobility is the key element in pastoral strategies in all parts of savanna Africa.[12] Pastoral movements are far from random. They are not only based on seasonally shifting ecological conditions, but also, as Stenning said as long ago as 1959 of the Fulani in Nigeria, 'on an intimate knowledge of a tract of country, its human, bovine and animal population, its resources in pasture and its marketing possibilities'.[13]

Variations in the timing, and the unpredictablility, of rains present significant problems to pastoralists. Pastoralists cope with such problems using both indigenous knowledge of the environment, and through kinship and cattle sharing and exchange networks. The ability of the Gabbra of northern Kenya, for example, to survive droughts depends on their skill in interpreting history (and their belief in the cyclical return of events) and their knowledge of genealogy and relationships that enable them to call for help when stock die. Paul Robinson comments that 'those Gabbra herd owners who consistently emerge from crisis situations with the greatest percentage of their herds intact are precisely those who most closely recognise and follow the Gabbra cycles'.[14]

Knowledge of environment, environmental change and landscape is a vital element in the success of pastoral strategies. There are, however, numerous others.[15] For example, many pastoral groups run mixed herds of sheep, goats, camels and cattle. Mixed herds may not look good to pastoral experts trained in northern temperate areas, but they make sense in Africa. Different animals can exploit different elements within the available vegetation. In this way the available food resource is better utilized. Goats and camels are browsers, and hence exploit the woody biomass of trees and shrubs. Such plants are deep-rooted and get access to shallow groundwater. They go on growing for far longer than grass. Cattle are dependent on grass which is in turn wholly dependent on the annual rains.

Not only are flocks and herds mixed, they are also structured in terms of age and sex in very particular ways. Herds have a dominance of female animals (between two-thirds and three-quarters of the herd would be typical).[16] This is because of the importance of milk rather than meat as the main pastoral product. Pastoralists also depend on non-pastoral foods, obtained both through gathering and through trade with farmers.

The mixture of livestock and food sources, combined with mobility, are the basis of the flexibility and resilience of pastoral systems. The Turkana of northern Kenya, for example, obtain the most important element in their diet, milk, through two quite separate pathways. The first is from camels, which feed on shrub vegetation that remains green well beyond the end of the rains. The other is through cattle which depend on grass which is only available for a short period after the rains. Camels lactate through the year, while cattle are very productive of milk but only for a short period. Cattle are also more restricted in where they can graze because of their need of water. The use of the two pathways thus provides a secure basic nutritional supply through the year, while also making use of seasonally available production. In the long dry periods, Turkana households split and the men take the cattle away to the hills. Goats and camels are kept on the plains.[17]

The flexibility of African pastoralists is an important feature of indigenous responses to seasonality and drought. The history of their relations with the environment provides evidence of this flexibility over long time periods. The effect of the rinderpest epidemic in East Africa is a good example of this flexibility in action. Rinderpest is a disease that kills cattle and wild ruminant herbivores like buffalo. In the 1890s, it entered North-east Africa, and ran rapidly south.[18] It decimated populations of wild herbivores and, more seriously, destroyed cattle herds. In Maasailand it struck early in 1891, coming on top of an outbreak of bovine pleuropneumonia a few years before. By the end of 1891 the Maasai faced famine, and warfare broke out between Maasai sections.[19] These disasters were followed by a smallpox epidemic in 1892 and drought in 1897. Into the vacuum created moved the German and British colonial powers.

As a result of these various calamities, part of Maasailand, the Mara Plains in south-west Kenya, were depopulated in the 1890s. As a result of reduced cattle grazing, scrub began to encroach on the grasslands, and with the scrub spread Tsetse fly and sleeping sickness. The Maasai who began to move back into the area from 1900 onwards were strangers to it, from tribal sections that had best survived the perils of disease and the translocations forced by colonial government. They used the grazing resource very differently and less intensively, largely avoiding the fly-infested scrub. Tsetse in fact continued to advance until large-scale bush clearance in the 1950s. Richard Waller comments that Maasai pastoralism had 'a high degree of resilience and adaptability and has been able to cope successfully with a range of changes and pressures, including colonial intervention'. He also notes that in the process it underwent considerable modification.[20]

Flexibility on the part of African pastoralists also involves the use of the full range of environments open to them, both within the rural environment and outside it in urban areas and (temporarily) in relief camps.[21] In Turkana, areas of woodland along seasonal rivers are vital to survival of herds through the dry season. In West Africa, both large and small wetland areas along major rivers provide an essential dry-season grazing resource. Such areas represent small proportions of the total area exploited by pastoralists, but they have an importance which is disproportionate to their area. These special areas have often been the target of attempts to 'develop' semi-arid regions, often (as in the case of irrigation in northern Kenya) in the name of sedentarisation and 'development' of pastoralists themselves. The implications of such development for indigenous use are discussed in later chapters.

Such development represents just one end of a continuum of interventions by outsiders in African pastoral societies. With few exceptions, such interventions have narrowed the resource base of pastoral groups and reduced their freedom of action. Rural development projects in general, and pastoral development projects in particular, have been influenced by a long intellectual tradition of anti-nomadism. The Commissioner of the East African Protectorate, Sir Charles Elliot, wrote of the Maasai in 1903 'morally and economically they seem all bad'.[22]

It is now widely recognized that the unthinking transfer of ideas about pastoral development based on ranching technology have been not just inappropriate but also damaging.[23] Inappropriate technologies, irrelevance to indigenous economies and needs, high capital costs and centralized management have rendered many livestock development project failures. At the same time, pastoralists themselves have faced numerous problems, including conflicts with other land users over resources, loss of key grazing lands to development projects (particularly irrigation schemes) and the transfer of livestock capital from the hands of subsistence pastoralists to rich urban investors.

## Desertification

The perception of African pastoralism that has probably had the most significant impact on policy during this century is that it causes the degradation of semi-arid environments.[24] A recent review commented that 'for decades, most pastoral development projects in Africa have been based on the principle that self-interest makes pastoralism underproductive and environmentally damaging'.[25] Such perceptions were greatly boosted by the Sahel drought of 1972–74, and the famine that accompanied it. These events shocked a complacent First World public. They placed the vagaries and extremes of Africa's climate on Northern television screens and the global political agenda. International and non-governmental relief agencies had been warning of famine for months before the media and First World politicians took the threat seriously. In the mid to late 1970s the drought eased (or at least changed: the complexities of what has actually happened are discussed below), and African environment and development rapidly dropped from the international agenda.

When drought and warfare again drove the Sahel and the Horn of Africa into famine ten years later, the same cycle was repeated. Again, although the agencies on the ground were able to predict the coming disaster well ahead, in the UK it was only when a BBC reporter's film and account of the plight of refugees reached the evening television news that the machine lumbered once more into action, this time associated with the astonishing response of young people in many countries in the Band Aid concert. Once again the hazardous environment of Africa and the problems of poverty, population growth and war became familiar topics of public debate in the First World. But media and political memories are short, and the spectre of famine in the Sudan and Ethiopia in 1990–91 again met with a sluggish response from First World governments.

Twice in the last twenty years, international concern about Africa has boomed and bust. Outside observers of Africa, commentators and many researchers, have portrayed Africa as a continent locked in crisis; economic, political and, above all perhaps, environmental crisis.[26] There is widespread concern about the degradation of arid lands under rising pressures of use by people and livestock (falling fallow intervals, rising populations, soil erosion and loss of vegetation cover) and associated problems of fuelwood shortage in the semi-arid areas of Africa and the rapid rates of deforestation in the moist forest zones, both for timber and agriculture.[27] These concerns are very often expressed in apocalyptic terms. Such writing about Africa in the 1970s and 1980s is part of a long tradition.

Concern about land degradation and the spread of deserts on Africa has been expressed for over a century, especially in southern Africa.[28] In 1934 E.P. Stebbing, Professor of Forestry at the University of Edinburgh, visited West Africa. The aridity of the Sudan/Sahel savanna in Northern Nigeria and the intensity of its use led him to argue that there was a process of 'progressive desiccation'. He believed that the area was drying out and the Sahara was advancing southwards and he promoted this idea in a number of publications in the 1930s.[29] Other writers argued that he was wrong. In 1938 Brynmor Jones reported that the Anglo-French Forestry Commission which visited the area 1936–37 had found no sign of large-scale encroachment of sand, no retrogression of vegetation, observed reduction in rainfall or fall in water tables. Jones maintained that local soil erosion could be dealt with 'without resort to expensive regional schemes'.[30] Debate about soil erosion and desiccation was by no means restricted to Africa. The experience of the US dust bowl influenced perceptions about semi-arid lands in many countries through the 1930s and 1940s. The resulting concern was global in scope, and is recorded in books like *L'Erosion du Sol* in 1947, by the French geologist Raymond Furon.[31]

Government agriculturalists in Kenya, influenced by knowledge of the American dust-bowl and faced with coincident conditions of drought, the effects of world depression on the white settler agricultural economy and a rise in squatter populations on settler land concluded that there was a serious problem of environmental degradation.[32] Their response was just the kind of major intervention that Jones hoped to avoid in Nigeria. Extremely unpopular programmes of terrace construction were introduced in Kenya and in other parts of anglophone Africa. Meanwhile, in Nigeria, echoes of Stebbing's concerns lingered on through the end of the colonial period and into independence. They were awakened by the droughts of the 1970s and 1980s and in the end provided one of the justifications for government investment in large-scale irrigation through the river basin authorities in the 1970s.[33] Such ideas are still major elements within official thinking about environment and development in the north of the country.

Overviews of the droughts and famines in Africa in the 1970s and 1980s have tended to focus on the problem of 'desertification'. The word was coined by the French ecologist Aubréville in 1949 to mean the process by which desert-like conditions (arid areas with few plants) develop.[34] In the 1970s use of the word expanded enormously, and desertification became a central element in debates about Africa's 'crisis'.[35] By the late 1980s it was clear that there were significant problems emerging in the usage of the term. It was ambiguous and had suffered what one analysis wryly described as 'an erosion of meaning'.[36] It was widely taken to mean any loss of biological productivity which might lead to desert-like conditions. Thus this 'desertification' included waterlogging and salinization of irrigated land as well as loss of biomass, loss of vegetation cover or soil erosion.[37]

Part of this verbal inflation can be traced back to the United Nations Conference on Desertification which was organized by the United Nations Environment Programme in Nairobi in 1976. Worthy though this effort was, it made 'desertification' a highly politicized concept. Aid agencies wanted to be seen to put money into stopping it, and African governments saw it as a way to obtain aid. Through the 1970s and 1980s existing development and aid programmes in arid areas were re-labelled 'desertification control' projects (much in the same way as European governments discovered 'green' projects in their portfolio in the late 1980s). However, the amount of innovative work either in terms of research or practical action was limited.[38] In particular, there was very little scientific study of what phenomena were occurring and where, and whether they were getting worse or better.[39]

New technology such as satellite remote sensing has generated more data, but raised new questions. It is possible, for example, to derive an index of the greenness of vegetation by calculating the ratio of radiation reflected from the earth's surface in different wavebands using a satellite-mounted sensor. From this, primary production over large areas can be calculated.[40] Comparison of data from the National Oceanographic and Aeronautical Administration's Advanced Very High Resolution Radiometer for successive years in the early 1980s allowed differences in vegetation cover of the Sahel to be estimated.[41] This provided a graphic visual expression of the ecological impacts of the early 1980s drought. However, what did this mean in terms of the long-term productivity of the ecosystems of the Sahel? To what extent was this evidence of land degradation? What implications did it have for agriculture or pastoralism? More importantly, what did it imply for the complex social and economic systems (for example reciprocal trade between pastoralists and farmers) on the ground? These questions were harder to answer.

It is now clear that simple and rather sweeping ideas about desertification are unhelpful, and can be seriously misleading. For example, there is remarkably little good evidence to support the widely-held view that the Sahara is advancing southwards year by year. Research in the 1970s on the Sudan suggested that this was the case, on the basis of a comparison of 1958 vegetation maps and 1970s satellite imagery. This study is widely quoted, but Ulf Helldén showed in 1988 that the two surveys are not strictly comparable. His own studies of central Kordofan sugested that it was not possible to show the systematic advance of the Saharan boundary, nor the encroachment by active sand dunes. By contrast, there had been significant northward expansion of the limit of cultivation over the previous century. The 1964–74 drought greatly affected the productivity of both natural vegetation and crops, but there was a rapid recovery as the drought ended.[42] This study only refers to one area, and many such studies are needed to establish the dynamics of African ecosystems in the face of population growth, changing economies and drought. Without such studies, generalizations about processes of desertification and the rates at which these processes are happening, let alone their causes (natural or human) must remain an unreliable guide for policy makers.

Most commentators now seem to prefer to discuss land degradation rather than desertification.[43] This concept embraces a wider range of ecological processes involving reduced productivity. Thus Nick Abel and Piers Blaikie define rangeland degradation as 'an effectively permanent decline in the rate at which land yields livestock products under a given system of management'.[44] Such a concept of land degradation makes it specific to the management system in force, and implies that natural processes will not rehabilitate the land and that it is not economically worth investing in rehabilitation projects.

Land degradation, as defined in this kind of way, is without doubt a serious problem in sub-Saharan Africa, but it takes place in patches where local land use is unsustainable. The most serious damage in terms of soil erosion or loss of productivity occurs well away from the sub-Saharan zone in the wetter savanna, where there are more rainfall, more agriculture and more people. It is argued that the popular view of Sahelian livestock populations in decline is not supported by the data, which show increases over the last two and a half decades of between 20 and 100 per cent.[45] It is also argued that agricultural production has risen steadily in parts, at least, of the Sahel (although notably not the Horn of Africa), while the area of cultivation has increased by about one third. Now a lot of these statistics are in the long run very bad news for rural Africans, but they do suggest that common perceptions of Africa's environment are unreliable as a basis for future action. It is clear that past predictions of environmental disaster in the Sahel may have been premature.

It is remarkably difficult to make entirely satisfactory statements about environmental change on the ground in Africa.[46] There are still too few quantitative field studies of land degradation in Africa. Many studies have been essentially experimental, and are not being carried on into regular programmes of monitoring. Many techniques like remote sensing are relatively new, complex and costly. Few African countries have the computer technology necessary to even begin handling satellite data themselves. There are simple technologies (not least among them simply talking with farmers), but these are labour-intensive and complex to administer. They have their own costs, and relatively little aid agency money has flowed to support such programmes.

Above all, however, most data on the environmental conditions (not least rainfall) are short and cover small areas. Yet the variability of African climates is such that if there is to be any chance of identifying environmental trends, data sets need to cover long periods and large areas. Without long series of data, it is impossible to separate out the impacts of individual drought years (during which one might expect grass cover, for example, to be much reduced) from longer-term systematic degradation caused by overgrazing. It is equally impossible to determine what 'normal' conditions would be like, so that deviations from it could be determined. What year would make a sound baseline in the Sahel, for example? Somewhere in the 1960s, when rainfall was higher than average? The 1970s, during or after the drought? The 1980s, when rainfall patterns were seen to be complicated in space and time? Even where scientific data exist, interpretation may be difficult. Serious degradation to pasture land can occur without loss of biomass: the invasion of scrub or unpalatable grass species can have serious effects on the ability of pasture to support livestock, for example, without affecting total plant biomass at all.

The biomass of pastures may also be severely depleted without any long-term damage, for example in isolated drought years. Indeed, savanna pastures are subject to a severe seasonality in biomass production every year. To a visitor from a temperate Northern country most of the Sahel resembles a desert by the end of the dry season in even good years, simply because most of the available pasture resource has been removed. Some is eaten by wild animals, a large part is eaten by termites, and some dries out and blows away. A good fraction is consumed by domestic livestock and converted into meat and milk on the hoof. The critical feature of this annual cycle is the extent to which such pastures recover during the rains. Andrew Warren and Clive Agnew point to this as the key factor: the resilience of semi-arid ecosystems. After all, the savannas of Africa have evolved under conditions of seasonal aridity. The species within them are adapted to avoiding or recovering from desiccated conditions. Savanna plants survive defoliation and seasonal drought, especialy annual grasses and forbs. They also survive years of low rainfall and bounce back. The drought in Kordofan in the Sudan in the mid-1980s was so severe that even local pastoralists were concerned whether pastures would recover. The rains of 1988 were exceptionally heavy, causing widespread havoc (and among other things flooding Khartoum). By early 1989 annual pastures on sandy soils had recovered completely, although those on cracking clay soils had not.[47]

Under some conditions of human land use, savanna plants may lose that capacity to recover. That is the point at which serious land degradation is occurring. It is a point of great practical importance to African farmers and pastoralists, for it determines when poverty starts to slide into destitution. However, it is scientifically difficult to determine. The task is made harder by the confusion over the term 'desertification' and the rhetoric associated with the problem among policy makers and politicians.

## Changing climate

Seasonality and within-season variability in rainfall are, in a sense, part of the everyday experience of cultivators and pastoralists in Africa. There are, however, longer-term patterns of rainfall which are harder to explain, and tend to be labelled as 'disasters' when they occur. The most obvious of these is the phenomenon of 'drought' years, that is years with unusually low rainfall.

The variability of annual rainfall increases as total annual rainfall decreases. This means that the variation between years is proportionately greater in the drier parts of Africa such as the Sahel. Of course, in higher rainfall areas the absolute changes from year to year can be quite large, but they tend not to be noticed. Where rainfall totals are low, and many economic activities are marginal, small absolute changes in rainfall can have a major and sometimes disastrous effect. Some areas such as Turkana in northern Kenya experience major differences in rainfall from one year to the next, with some years receiving only 40 per cent of the mean.

The 'drought' of the early 1970s in the Sahel, a string of rainfall deficit years, began in 1968 and peaked in 1972–74. Some would argue that it never really ended, since rainfall totals in much of the Sahel remained below the mean values of the first half of the century through the 1970s, reaching a new low in the early 1980s (see Figure 3.2).[48] The rainfall deficits in the 1970s and 1980s have not been uniform through the season. The greatest percentage reduction in rainfall has been in August, particularly in the Western Sahel.[49] This is the wettest month in the Sahel, during which the seed of grain crops usually fills out.

The 1970s drought followed a decade that was wetter than the 1941–1970 average. It came as a surprise to some in the development community, but it was not without precedent. There have been other dry periods, such as the 1940s and the 1910s. The geographer A.T. Grove presented a paper in Addis Ababa in the early 1970s which described the drought and famine in northern Nigeria in 1913 which followed a series of low-rainfall years from 1898 onwards.[50] At this time, the level of Lake Chad was low, and the northern lobe had dried out (as it did again in the early 1980s). The discharge of the Nile was also lower than it had been for centuries. Indeed, the whole period of the early twentieth century was similar to the conditions of the last twenty years, dry in the western Sahel, with an intense but shorter drought in the central Sahel.

It is not very useful to talk in general terms about drought in Africa. Recent research stresses both the variability of rainfall between adjacent areas, and also the need to be specific about what drought means. Work in Niger, for example, distinguished between 'meteorological drought' (i.e. a change in precipitation) and 'agricultural drought', meaning rainfall shortfalls that actually lead to failure of crops, specifically of millet. In the 1970s both kinds of drought were common in the drier North of the country (with less than 500mm rainfall). Further south, where rainfall is higher, meteorological drought was far less common in the 1970s and there was little evidence for agricultural drought.[51]

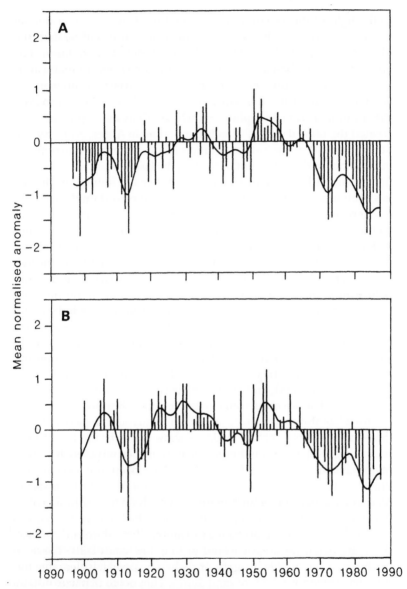

*Notes:* A. Western Sahel (10–15° N, 20° W–10° E)
B. Central Sahel (10–15° N, 10–40° E)
The data show the mean rainfall anomaly in each year in the two regions compared to a reference period of 1941–70. The continuous line shows a rolling ten year mean. The figure is redrawn from G.Farmer (1989), 'Rainfall', pp. 1–16 in *The IUCN Sahel Studies*, International Union for the Conservation of Nature, Gland, Switzerland.

**Figure 3.2** Annual variability of rainfall in the Sahel

In the light of this variation in precipitation, how sensible is it to see the first sixty years of this century as 'normal' in terms of rainfall? It is, of course, the period of widespread rainfall records, and thus it is the only quantitative baseline against which more recent rainfall can be assessed. It is also the period over which expatriate Europeans have been observing Africa. It has thus represented 'normality' for many of the administrators and planners whose ideas and opinions have created the official view of what is happening in Africa, and who have determined the institutional memory of the modern urbanised bureaucrats of the continent. However, the last 60 years is a misleading time-frame within which to view African climate. Rainfall in the past has been much more variable than the recent record might suggest.

There are few meteorological records before the end of the nineteenth century. However, there are historical sources such as accounts of famines in the kingdoms of the Niger bend from which Sharon Nicholson has compiled a historical account of climatic variations in recent centuries in West Africa.[52] Runs of dry years have occurred before on a number of occasions. There were famines in Borno and in the Niger bend in the eighteenth century, and an arid period seems to have lasted from about 1680 to the 1830s. There is also a remarkable record of the level of the Nile recorded at the Roda Nilometer at Cairo from AD 641. This is hard to interpret, but again seems to show periods of high flows in a series of years (e.g. 1351–1470, 1737–1770, 1850–1900 and 1950–1974), and periods of low flows. It seems that droughts have been a feature of African climate for many hundreds of years.

Africa has also experienced significant climatic change over longer time-scales. Various sorts of evidence show that over the last 250 centuries (in geological terms the late Quaternary period) there have been major climatic changes in Africa. There have been long periods both much wetter and much drier than the twentieth century. In various places on the margins of the Sahara (especially on the north side of the Niger Inland Delta in Mali, around Lake Chad and in the western Sudan),[53] and on Kalahari Sands in Zambia, Zimbabwe and Angola[54] there are extensive areas covered in long low sandy hills. These are easily visible on satellite imagery, and when mapped it is clear that they were once extensive fields of sand dunes. They occur in places now too wet for sand to be blown by the wind, except in the driest years. Those in the 'ancient erg of Hausland' were active 12,500–20,000 years ago, at which time it must have been considerably drier in these areas.[55] In the late Quaternary, dunes obstructed the River Niger itself upstream of Timbuctou, and the Senegal at Kaedi.[56] Around 18,000 years ago, vegetation belts lay south of present positions along most of the length of the southern Saharan boundary. Around 8,500 years ago, vegetation belts lay some 4–500 kilometres north of their present-day positions.[57]

Other evidence demonstrates that conditions have also been wetter than today, particularly between 12,500 and 4,500 BP. Pollen remains in the remote Selima Oasis in the desert of northwest Sudan suggest that thorn savanna was present between 8,400 and 6,000 years ago.[58] Lake shorelines have been discovered and mapped at altitudes well above present-day lake levels, notably in the East African Rift.[59] A former high point of Lake Chad has left a remarkable landscape feature across northeast Nigeria in the 12 m high 'Bama Ridge'. This now lies several hundred kilometres back from the 1970s position of the lakeshore. The enlarged Lake Chad (called Mega Chad) covered 330,000 square kilometres. In some cases it is possible to use calculations based on the former volume of the lake and the area of the river basins feeding it to estimate former rainfall conditions (although this requires numerous estimates and assumptions of likely vegetation cover and evaporation).

The evidence from lake levels shows that conditions were substantially wetter after the end of the last glacial advance in the northern hemisphere. Wetness began to increase about 12,000 years ago in West Africa. By 9,000 years ago, wetter conditions were widespread in West Africa, peaking at about 8,500 BP.[60] This evidence of wetter conditions is supported by discoveries of rock paintings in the Saharan massifs such as Tibesti of animal species now locally extinct (e.g. hippo and elephant), from the discovery of fossils of species such as elephant, crocodile and fish such as Nile perch in desert sediments in Ahaggar and Tibesti, and by the patterns of distribution of families of freshwater fish and other organisms. Rainfall must have been almost 10 per cent greater than in the twentieth century. Flow in the Logone/Chari system (and perhaps in the Niger and Senegal rivers as well) was five times greater, and many valleys feeding into the northern flank of the Niger in Niger that are now permanently dry carried flow at this time.[61]

From about 8,000 years ago conditions grew drier, and lakes shrank again. Lake Chad appears to have been dry between 4,500 and 3,700 years ago. From between 4,000 and 2,000 years ago wetter conditions again prevailed and vegetation patterns broadly similar to those of today were established.[62] In recent years, a great deal of new evidence has accumulated from the analysis of sediments and cores from the ocean bed off Africa. New sources are building up a more complete picture of climatic change in recent geological times, but are not changing the basic fact of extensive climatic variation over the last 20,000 years or so. Over such time-scales, dry periods were both longer and more intense than any experienced within recent centuries. This is important, because it is just in this period that systems of herding and agriculture were developing in tropical Africa. Over more than historical time, drought has been part of the common experience of African farmers and herders. The 'droughts' of the twentieth century, serious though they are today, would be minimal blips on the graph of longer-term climatic change.

Some researchers have tried to identify patterns in the occurrence of wet and dry years in Africa. This has been done quite successfully in southern Africa,[63] but has proved less easy in the Sahel. Similarly, a number of theories have been proposed to explain the prolonged post-1968 drought. Some embrace local effects such as changes in surface conditions, conceivably because of human action. Others try to link Sahelian rainfall with surface/atmosphere interactions on a wider scale.

Among the most widely quoted mechanisms is the 'albedo hypothesis'.[64] Loss of vegetation cover in an area like the Sahel (e.g. through grazing or woodcutting) could increase the reflectivity of the land surface and hence affect heat exchange between earth and atmosphere, in such a way that uplift and subsequent rainfall were reduced. Similarly, a drier surface would yield less evaporation, give a changed surface energy balance and less convection and hence rainfall. Both 'biogeophysical feedback' mechanisms have, in theory, the capacity for 'positive feedback', in that once begun (e.g. through human action) their effect could progressively intensify. A few researchers have made observations of conditions in the field and a rather larger number have undertaken complex computer modelling exercises to investigate these mechanisms. The computer experiments do show some response of rainfall to changed surface conditions, but only if start conditions are made far more dramatic than those actually observed in Africa.[65]

Other studies have linked changes in rainfall in Africa with larger-scale changes in oceanic and atmospheric conditions. Among the most interesting is the apparent link between rainfall in the Sahel and sea surface temperatures, particularly in the Atlantic. There is a statistically significant linkage between low Sahel rainfall, warm sea surface temperatures south of the Equator and cold temperatures to the north.[66] It now seems clear that droughts in the Sahel in the last 14,000 years have been associated with injections of fresh water into the western Atlantic.[67] This work is an important reminder that neither the Sahel nor Africa as a whole is isolated from the global atmospheric system. Larger-scale influences on global atmosphere and climate, from changes in the geometry of the earth's orbit to the 'greenhouse effect' are therefore likely to be important to African climate. The causes of rainfall variability in Africa remain somewhat uncertain, as does its annual pattern. What is certain, however, is that variation is a fact. Furthermore, it is a fact that both African land users and those planning development projects have to live with.

## Measuring the rain

Africans have lived with the unpredictability of African rainfall for millenia and have developed a series of neat and more or less effective ways of insuring themselves against the vagaries of climate, while exploiting the natural resources available to them. Industrialized 'northern' humans have not been in Africa so long, and the reality of droughts has tended to startle and alarm them. British colonial officers in northern Nigeria faced with the drought and famine of 1913 not only did not know what to do about it, they also did not understand what was happening. To someone judging semi-arid Africa by the standards of what is normal in a temperate environment, the variability of climate is both bizarre and incomprehensible. Ideas of 'normal' rainfall, based upon notions of 'average' rainfall conditions, might be appropriate to temperate environments, but have tended to prove a poor basis for planning in Africa. It could be argued that much of the psychosis about 'desertification' in anglophone Africa stems from this basic alarm, both transmitted through the institutional memory of African bureaucracies and re-experienced by successive generations of administrators and 'experts'.

Not only has African rainfall surprised those coming in from outside, it also tends to find flaws in their attempts at scientific understanding. Development planning is served by technical experts with a formidable array of sophisticated scientific methods. Unfortunately, such methods, and the related computer hardware and software and textbooks, are mostly developed in temperate climates of Europe or North America. Neither these environments, nor those of the arid American southwest or the USSR, provide an adequate analogue for the variability of African rainfall. Hydrologists and engineers have nothing like the same grip on the patterns of nature in Africa as they have been trained to expect. The use of computer simulation and statistical analysis serves to give confidence to the modern development planning process, but it does not always succeed in comprehending the variability of the African environment.

The variability of rainfall within a single wet season presents serious problems for development planning. As has been remarked upon above, there are few rainfall stations in semi-arid Africa, and most do not have long records. Many former rainfall stations have fallen into disuse, some in the last few years. Ironically, data are particularly scarce in areas such as the Sahel which are particularly critical. When researchers wanted to examine year-to-year variations in Sahelian rainfall in the mid-1980s they were only able to find five stations with a record as long as 50 years.[68] There were many other stations with shorter, or incomplete, records, and a considerable amount of statistical juggling was required to derive usable data which they could argue could represent the whole Sahel. Without long records it is extremely difficult to establish any kind of figure for 'normal' rainfall against which the impact of drought years can be assessed.

Furthermore, it is customary for maps of annual or monthly rainfall to be drawn simply by 'joining the dots' between widely-dispersed rainfall stations. While this may give a useful overall impression across the continent, at any finer scale it is pretty meaningless. The pattern of rainfall identified by a consultant designing a development project is likely to be very different from the reality experienced by a farmer planting and managing crops.

Data on river discharges suffer much the same problems as those of rainfall, perhaps even more severely. Many gauging stations have disappeared. Many others have not been maintained. Gauging stations need to be re-surveyed every year or so to re-establish the mathematical relationship between water depth and discharge. This requires measuring the cross-sectional area of the river and water velocity over a period. On a large river in flood this requires a bridge, a boat or a wire cradle so that a surveyor can dangle above the middle of the river. Without this periodic re-measurement, water level figures give an increasingly inaccurate picture of actual flows. These problems are particularly severe in river channels which are sandy and which change a lot from year to year because of erosion or deposition. Many rivers in semi-arid Africa are like this.

Many hydrological departments cannot even find the money to train and pay recorders to measure water depth, let alone the money to maintain such equipment and to transport qualified surveyors into the field. By and large, the maintenance of a hydro-meteorological network has not ranked high on the priorities of aid donors. Even once collected, hydrological data are often badly organized and preserved in African countries. Sometimes they are physically deteriorating (e.g. in books being eaten by termites or damaged by water). Sometimes data have been lost. In one country, I eventually traced river discharge data to three filing cabinets in an annexe of a Ministry of Sport and Culture. On another occasion, a colleague and I were allowed to take original hand-written hydrological yearbooks away to photocopy: a generous act, but hardly sensible practice.

Rainfall records are usually available over a much longer period than river discharge figures. This is not surprising, since they are much cheaper and easier to collect. Hydrologists conventionally 'stretch' available river flow records by correlating them with longer rainfall series and by regional studies of rainfall and runoff.[69] In this way it is possible to generate a 'simulated' river flow series. This can in its turn be processed to estimate the size of floods which are likely to recur every few decades or centuries (to make sure dams and other structures are strong enough), and to estimate low-flow years to make sure that water storage is sufficient to cope with droughts. Many projects lack specific data, certainly for any long period, and hydrologists have to use data from upstream and downstream and regional data from analogous areas in their calculations. It is not impossible to find development projects costing hundreds of millions of dollars based on less than ten years of monthly average flow data, and perhaps only one or two years of daily flow data. Predicting the magnitude of the flood or drought likely to recur once in a hundred years from such data sets is extremely hazardous.

Where there are no discharge figures at all, there is a method of making an estimate using rainfall data and some standard figures for the mathematical relationship between rainfall and runoff. This is based on actual measurements of the effect of different types of soil and vegetation cover on the proportion of rainfall which runs over the surface (or though surface layers) and gets into stream and river channels. This will vary with the nature of the ground surface (particularly the vegetation cover), and also the pattern of the rainfall (for example the intensity and duration of storms). Estimates using this approach are usually pretty crude, although they may be useful in some circumstances. In Africa, relatively few actual measurements of rainfall-runoff relationships have actually been made, so figures are often taken from somewhere which is better-studied. This may be an adjacent region, or another part of Africa. I have known data to be used from the southwest USA. This does not improve the predictive accuracy of the method. Lack of local rainfall data exacerbates the problem of making methods such as this yield reliable estimates of river flows.

The drought years of the 1970s and 1980s have had major impacts on river flows and the status of wetlands in Africa. In 1983 the aluminium smelter at Tema in Ghana shut down because of lack of flow in the Volta. Power supplies for the dams at Kosson in the Ivory Coast and Kainji in Nigeria were reduced. In the 1984–5 dry season the Niger at Niamey was at a record low. Flows in the Senegal were also low in the mid-1980s, and there was a serious problem of die-back on coastal mangroves due to salt water intrusion.

More dramatically, perhaps, in the 1980s Lake Chad shrank to a fraction of its size twenty years before. For much of this century, Lake Chad has covered about 20,000 km$^2$, and had a volume of about 80 cu. km. At least half of the inflow came from the Logone-Chari system in Cameroon. The lake level varies annually, reaching its lowest just before the rains in April–June. Through the 1970s the levels fell, and the lake progressively shrank. It recovered slightly in the late 1970s, then shrank again below previously recorded minimum levels. The northern lobe of the lake dried up altogether. In 1984–5 the inflow from the Logone-Chari was so slight that the lake level did not rise, and the lake covered less than 10 km$^2$ (see Figure 3.3). Inflows have remained small, and most of the lake was still dry in 1990.

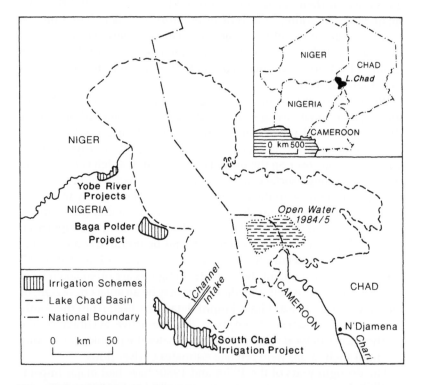

**Figure 3.3**  The Lake Chad Region

At the same time, rivers in other parts of Africa have experienced equally unexpected increases in discharge. The discharge of the Zambezi between 1948 and 1980 was about 40 per cent higher than that between 1914 and 1948, and there were significant problems from high flows during the construction of the Kariba Dam.[70] In the 1960s, above average rainfall raised the level of Lake Victoria and the discharge of the White Nile. From 1962 to 1980 discharge of the White Nile into the Sudd was about 40 billion m³, double that experienced in the first half of the century. These high flows effectively compensated for the reduced flows in the Blue Nile, draining the drought-stricken Ethiopian Highlands. In the 1980s discharge in the White Nile fell, coinciding with renewed drought in Ethiopia, with significant effects on storage in Lake Nasser.[71] In 1984 the discharge of the Nile into Lake Nasser (behind the Aswan Dam) was less than 60 cu. km, compared with the average flow of 84 cu. km between 1900 and 1959.

## Implications for development

The variability of rainfall in Africa obviously has serious implications for development planning. It is easy to be wise in retrospect, but the climate of Africa in just the last two decades gives little comfort to those trying to develop water resources. Currently, we can neither completely explain the droughts like those in the Sahel in the 1970s and 1980s, nor can we predict whether they will persist, or (if they ease) whether and when they will recur. We cannot say whether they represent only some kind of natural variation, of the kind which has happened many times in past centuries and millenia, or whether they are in some way related to human action, and hence population growth and land use change. Both possibilities offer a bleak future; there is little prospect of major shifts in patterns or intensities of land-use, and populations continue to soar; there is no prospect currently of influencing natural climatic variations in a controlled and beneficial way.

Published rainfall and river flow data are often out of date. Graham Farmer, writing in 1986, comments that:

> it would be ludicrous, for example, to use rainfall statistics for West Africa that did not incorporate the last 15 years, and yet most published reference works do not include those data.[72]

He points out that none of the annual rainfall totals for 1965–85 at Maradi in Niger reached the mean for the standard period 1931–60. Any calculations based on the earlier period would be extremely optimistic. Of course as time passes, data from the dry 1970s and 1980s are incorporated into the published record, and there are also technical aid projects aimed at improving the hydrometeorological networks in

Africa. However, to date those networks remain far from adequate. The problem – and the attendant risk – remains. Most major African rivers have now been dammed. Most of these dams have been built in the last thirty years, and most have been designed using very short runs of discharge data. There are serious problems in using such data sets for predictive purposes. It is extremely difficult to produce reliable estimates of the magnitude of the hundred-year flood for African rivers, just as it is to predict the drought likely to recur once in fifty years. Unfortunately, engineers and hydrologists do their best to make such predictions. The result is that very often there are unexpected impacts on downstream flows, and adverse impacts on the ecosystems and economic activities dependent on them. These are discussed in detail in Chapter 6. The Nile is remarkable among African rivers for the length of its discharge record, with gauged flows at Aswan from 1871 and written records of a sort from the Roda Nilometer from AD 871.[73] However, even with this length of record it is necessary to analyse data carefully to achieve adequate accuracy and consistency.[74] Less well studied rivers present much greater challenges to the ingenuity of hydrologists and developers.

Clearly, climatic variability should be an important factor in the planning of water resource projects in Africa. The variations in the level of Lake Chad mentioned above are of more than academic interest. Construction of a large scale irrigation project, the South Chad Irrigation Project (SCIP), began on the shores of Lake Chad in 1974. The project was to develop in stages, first 32,000 ha (gross area), then a final total area of 106,000 ha gross (67,000 net). The plan was to settle 55,000 farming families. Water for the project was to be drawn from Lake Chad via a 29 km long intake canal and raised by diesel pumps into irrigation canals. SCIP was commissioned in 1979, and a second project, Baga Polder, was begun in 1982. The fall in the level of Lake Chad from the 1972–4 drought onwards was a disaster for the project. In 1983–4 only 7,000 ha of crops were grown, and no crops were grown at all in 1984–5 or 1985–6.[75] Although the original supply canal was extended a further 24 km into the lake, in 1985 and 1986 the lake had shrunk to such an extent that the nearest water was 70 km away. Farmers (and indeed staff of the irrigation scheme) were busy growing crops on the dry lake floor using wells and small pumps, but by the mid-1980s the irrigation scheme was a disaster. The problems of water supply at SCIP have completely wiped out all predicted benefits of the scheme. The area irrigated never reached above 3 per cent of the predicted total between 1979 and 1986. The whole capital cost of the project has had to be borne with trivial returns from irrigated crops.

The hydrological studies done for the design of SCIP failed to predict the low lake levels of the 1980s. In retrospect, it can be seen that the data series available was too short for conventional hydrological methods to predict low levels. However, it reflects very clearly the limitations of such methods in African conditions. These limitations are further revealed by the fact that it is so difficult to predict when (if at all) higher levels in Lake Chad might return. The importance of the Logone-Chari inflow is so great that high rainfall totals in a single year in the headwaters of these rivers (in Cameroon and the Central African Republic) could have a dramatic effect on the area of the lake. This might have disastrous short-term effects on those now cultivating the lake floor. It is anybody's guess when, or if, this might happen.

Of course, it would be ridiculous to argue that development can only take place when perfect data is available: in Africa this would mean that almost nothing would get done. It does mean that data bases must be maintained and updated, and where data are limited or poor, great caution should be exercised by planners. It is the pride of engineers and hydrologists that they can come up with safe estimates where the data available are less than ideal. In this circumstance, much depends on the professional judgement and knowledge of the hydrologist, and the provision of a project design process which takes uncertainty into account. The lack of money and skills in African hydrological organizations means that such work is rarely possible 'in-house' by someone experienced in the region and its problems. The constraints of the consultancy business discussed above mean that there is seldom opportunity for visiting 'experts' to build up the kind of real practical expertise necessary to interpret the available data properly.

The environment of Africa is complex, unpredictable and often harsh. These are not facts which would surprise a Rendille herder in northern Kenya or an Ewe fisherman on the River Volta. They should not surprise those who presume to enter Africa to transform the environment in the name of 'development' of water resources, although, sadly, they often do.

# CHAPTER FOUR

# Using Africa's wetlands

*Poor people in a hazard-prone environment have a different kind of relationship with their land - it is less the relationship of a proprietor to property than that between rider and bicycle. (Paul Richards, 1988)* [1]

## Wetlands in dry lands [2]

On a dry season satellite image of West Africa taken by the American Landsat there are some remarkable splashes of red which immediately catch the eye. The picture, of course, does not have true colours. The Landsat image is created by a computer from a magnetic tape, transmitted down from the satellite to a ground receiving station, processed in various weird and wonderful ways, and projected onto a computer screen or a photographic plate. The satellite picks up the radiation reflected from the ground surface. Colours within the visible part of the electromagnetic spectrum (blue green and red) are allocated more or less arbitrarily to create an image we can see. It is customary to use red to represent reflectance in the red and infra-red part of the spectrum. On the ground, it is open water and green vegetation which reflects most within these wavebands.

So the red areas on the satellite image of West Africa represent water and lush vegetation on the land surface. These are wetlands, areas of seasonally flooded land set amidst drying savanna fields and pastures. They occur well to the north of the rainforest belt near the coast of Benin or Sierra Leone, deep in the Sahel. At ground level they present just as remarkable a picture as they do from space. One which I know quite well now is the Hadejia-Jama'are floodplain in northeast Nigeria. It is formed by the waters of the Hadejia and Jama'are rivers which flow northeast towards Lake Chad across flat dry country. This area receives some 6–700 mm of rainfall a year in a short 3–4 month rainy season. This supports a dry savanna grazed by Fulani pastoralists and rainfed farming based on bullrush millet. The rivers flood in the rains, and where they run into a field of old sand dunes, inactive for 15,000 years or so, they spread out into an intricate network of channels and pools.

This complex forms the wetland. Not all of it is wet by any means, for the dunes run through the area, but it is wet enough to support a remarkable range of economic activities. The general relief is slight, and the slope into the wetlands is imperceptible. However, there is a distinct ecological boundary between dry rainfed millet fields and rice fields, open water and irrigation. The boundary is particularly noticeable in the dry season, when the green fields of rice and beans planted on the drying land is in stark contrast to the arid landscape away from the water. Lakes and pools last well into the dry season, providing a fertile and watered place for flood-recession farming, grazing resources for large herds of cattle which arrive as the dry season draws on, and fish. These economic activities support a dense human population, and these people depend not on one but on the combination of the resources of the wetland. They exploit them with a skill and resourcefulness which is every bit as remarkable as the indigenous systems of resource management in the harsher drylands around them.

However, within semi-arid Africa, wetlands have a strategic importance out of all proportion to their size. In many instances, economic use of wetlands is integrated closely with that of surrounding drylands, and there are many examples of communities, with one foot in the wetland and another in the dry. For example, in Sierra Leone the cultivation of swamp rice and dryland crops are closely integrated. Dryland and wetland crops require labour at different times of year, and by exploiting the two environments, farmers are able to stretch over bottlenecks in labour supply, while at the same time making use of two separate environments and thus spreading risks.[3] Pastoralists also use wetlands seasonally, concentrating onto seasonally-flooded land as surrounding rangelands dry out. In this instance, a relatively small area of wetland provides support for grazing at critical times of year, and supports this activity through the rest of the year over a much larger area. For example, the Peul of the central Senegal Valley move away from the floodplain with their livestock in the wet season, but come back to farm when the floodwaters recede from the valley in the dry season.[4] Integration of valley and upland environments has been identified as one of the three basic features of indigenous agriculture in West Africa, along with physical management of soil and the practice of intercropping.[5]

## Indigenous irrigation

It is in agriculture that indigenous use of wetland resources in Africa is most intricate. Many different kinds of environments are exploited, using a range of techniques. Many of these comprise simple forms of irrigation. It is often thought that Africa does not have a tradition of irrigation. Engineers, in particular, are given to assuming that irrigation is an Asian or European technique, foreign to Africa's environment and people. Indeed, I have often heard that view developed in an overtly racist manner as justification for the failure of some large-scale modern irrigation scheme. Such attitudes appear to be common to both anglophone and francophone Africa.[6] There are various reasons for this blindness to irrigation in Africa. First, there has over time been remarkably little study of indigenous practices in wetland areas remote from centres of population and research. Second, engineers have in the past tended to adopt a rather strict definition of irrigation, to mean the controlled application of water to crops in a timely manner. Thus J.R. Rydzewski defines it as 'a special case of intensive agriculture in which technology intervenes to provide control for soil-moisture regime in the crop root zone'.[7] Much African wetland agriculture would be excluded by such a definition. However, it is now being realized that such a definition is unhelpful in the African context, and it is being recognised that in fact Africa has a strong and diverse tradition of irrigation, albeit of an informal kind and on a small scale.[8] Some of this is highly intensive, some is not, but all of it is characterised by John Sutton as 'specialised, ecologically sound, technologically complex and flexible'.[9] Yasmine Marzouk divides '*les techniques hydrauliques*' of Africa into three: first, soil moisture storage ('*le stockage de l'eau dans et sur le sol*'); second, the lifting of groundwater ('*les systèmes d'exhaure*'); and third, systems using gravity ('*les systèmes gravitaires*').[10]

In effect there is a continuum in African indigenous irrigation from simple adaptation to natural flood patterns in wetland areas to complete water control. At one end of the spectrum lie a range of techniques which involve use of natural flood patterns in floodplains or inland deltas, and techniques such as water harvesting (Table 4.1). At the other end lie systems such as the hill furrow irrigation systems of East Africa or the shadoof gardens of Nigeria or the sankiyas of the Sudan. These systems, which involve the creation of water-transfer and control structures which allow a high degree of control of water movement, have to be called irrigation by any standards.

Although it might be thought that the distinction between rainfed and irrigated farming is clear, in practice it is blurred, in Africa as elsewhere in the Third World. This is particularly the case outside the modern sector, and in the many places where farmers and other resource users have access to wetland environments. Increasingly, therefore, researchers and development agencies are now starting to adopt a much broader definition of irrigation to embrace wetland cropping systems well beyond the conventional concept of water application. The UK Working Group on Small-Scale Irrigation, for example, define small-scale irrigation as 'irrigation, usually on small plots, in which small farmers have the major controlling influence, and using a level of technology which the farmers can effectively operate and maintain'.[11]

**Table 4.1** A typology of African irrigation and water management techniques

A. *Flood cropping*

> rising flood cropping (planted before flood rises)
> *décrue* cropping (Residual Soil Moisture Cropping)
> > with bunds
> > without bunds
> flood/tide defence cropping (with bunds)
> > with freshwater inflow channels
> > without freshwater inflow channels

B. *Stream diversion*

> permanent stream diversion and canal supply
> storm spate diversion ('rainwater harvesting')

C. *Lift irrigation*

> from open water
> from groundwater
> > simple well
> > > –bucket
> > > –shadoof
> > > –animal-powered
> > > –motorized
> > tubewell

Debate about this is not simply splitting hairs about definitions, but has important implications for the way development agencies view indigenous systems of wetland management. It is the argument in this book that these wetland production systems are extremely important, and need to be taken into account in the development of wetland environments. 'Development' has to be something people do for themselves, not something imposed; it has to take account of the ways people use resources, and moreover how they use different *sets* of resources (e.g. wetland *and* dryland, or agriculture *and* fishing or pastoralism) rather than seeking to replace such activities with dramatic projects and transformation of the environment.

Probably the most surprising information about indigenous irrigation was that published in 1986 by the Investment Centre of the Food and Agriculture Organization (FAO).[12] This report was a desk survey of the extent of irrigation in forty countries in sub-Saharan Africa. It was compiled from published and unpublished data, and interviews with national and international specialists. The resulting figures should not be treated as if they were the result of careful field survey, because in some cases they are only guesstimates. The report itself comments dryly that 'the quality of the data reviewed was extremely variable and estimates for some countries were reached only by personal judgement after comparing irreconcilably conflicting sources'. Many previous studies had excluded flood cropping from consideration, but the FAO included it. Estimating the extent of cropping runs up against the problem of variation from year to year, and demands estimating 'average' flooded extent, and hence the area available for cropping. Given the variability of flood extent, especially through the 1970s and 1980s, this is a tricky task to attempt, and is unlikely to yield accurate figures.

**Table 4.2** The extent of 'small-scale and traditional' irrigation in sub-Saharan Africa

| Country | Area 'small-scale/traditional' ('000 ha) | Total irrigated area ('000 ha) | 'small-scale/traditional' as % total irrigated | |
|---|---|---|---|---|
| Nigeria | 800 | 850 | 94 | ** |
| Madagascar | 800 | 960 | 83 | ** |
| Tanzania | 115 | 140 | 82 | ** |
| Senegal | 70 | 100 | 70 | * |
| Mali | 60 | 160 | 37 | |
| Sierra Leone | 50 | 55 | 91 | ** |
| Sudan | 50 | 1750 | 3 | |
| Burundi | 50 | 52 | 96 | ** |
| Chad | 40 | 50 | 80 | ** |
| Somalia | 40 | 80 | 50 | * |
| Guinea | 30 | 45 | 67 | * |

| Country | Area 'small-scale/traditional' ('000 ha) | Total irrigated area ('000 ha) | 'small-scale/traditional' as % total irrigated | |
|---|---|---|---|---|
| Kenya | 28 | 49 | 57 | * |
| Burkina Faso | 20 | 29 | 69 | * |
| Gambia | 20 | 26 | 77 | ** |
| Mauritania | 20 | 23 | 87 | ** |
| Niger | 20 | 30 | 67 | * |
| Zaire | 20 | 24 | 83 | ** |
| Liberia | 16 | 19 | 84 | ** |
| Benin | 15 | 22 | 68 | * |
| Rwanda | 15 | 15 | 100 | ** |
| Botswana | 12 | 12 | 100 | ** |
| Ivory Coast | 10 | 52 | 19 | |
| Togo | 10 | 13 | 77 | ** |
| Angola | 10 | 10 | 100 | ** |
| Cameroon | 9 | 20 | 45 | * |
| Zambia | 6 | 16 | 37 | |
| Congo | 5 | 8 | 62 | * |
| Ethiopia | 5 | 87 | 6 | |
| Ghana | 5 | 10 | 50 | * |
| Mauritius | 5 | 14 | 36 | |
| Swaziland | 5 | 60 | 8 | |
| Malawi | 4 | 20 | 20 | |
| Mozambique | 4 | 70 | 6 | |
| Central African Republic | 4 | 4 | 100 | ** |
| Uganda | 3 | 12 | 25 | |
| Zimbabwe | 3 | 130 | 2 | |
| Lesotho | 1 | 1 | 100 | ** |
| Gabon | 1 | 1 | 100 | ** |
| TOTAL | 2.38 m ha | 5.02 m ha | 47 | |

Notes: 1.   **   75 per cent or more than of total irrigation in 'small-scale/traditional' sector

2.   *   50 per cent or more of total irrigation in 'small-scale/traditional' sector

3.   (Data not available for Guinea Bissau and Equatorial Guinea)

*Source:* FAO (1986)

Nonetheless, the data give pause for thought. What they show is that almost half the irrigated area in sub-Saharan Africa lies in what the

FAO call the 'small-scale and traditional' sector. The report identifies a total of 5.02 million ha of irrigation in sub-Saharan Africa, of which 2.38 m ha (47 per cent) are 'small-scale and traditional'. This category is diverse. It includes flood cropping, small earth dams, small run-of-river diversions, pump irrigation from wells or open water and water harvesting. Some of these (e.g most of the the flood cropping systems) will be of considerable antiquity, a number certainly pre-colonial in origin, and indigenous in both the sense of being initiated from within communities and also in the sense of being managed and controlled from within local communities. Others (e.g. the earth dams and pump systems) are more recent, and in many cases will be related to state programmes or even schemes controlled by a state irrigation bureaucracy. However, it is the indigenous irrigation which is the most important. FAO estimates that only 200,000 ha of the total 2.38 m ha is irrigated by what they call 'modern or intensive private operators'. The rest (92 per cent) is land irrigated by 'traditional private operators'. This gives some measure of the strength of indigenous irrigation in Africa. Given that almost no resources have been devoted to the support and development of indigenous irrigation, its extent is remarkable.

When the data are broken down by country, the picture is even more remarkable (Table 4.2). Taking all kinds of irrigation together, the countries with the largest area are the Sudan (1.75m ha), Madagascar (0.96 m ha) and Nigeria (0.85 m ha) (see Figure 4.1). Only four other countries have over 100,000 ha irrigated (Senegal, Mali, Tanzania and Zimbabwe). The irrigated area is only dominated by modern large-scale schemes in the Sudan (where 95 per cent of the irrigation is in the modern large-scale sector) and a few other countries such as Zimbabwe (which has quite a large area of large-scale private irrigation), Ethiopia, Swaziland, Mozambique and to an extent the Ivory Coast (Table 4.2). However the total irrigated area in most of these countries is small. In fact almost two-thirds (65 per cent) of all the 'modern' irrigation in Africa (both large- and medium-scale) is in one single country, the Sudan (1.7m ha).

Apart from the Sudan, most of the countries with fairly large irrigated areas are dominated by the 'small-scale/traditional' sector. Table 4.2 and Figure 4.1 show that in seventeen countries (out of thirty-eight) the 'small-scale/traditional' sector comprises more than 75 per cent of the total irrigated area, and that in a further nine it comprises more than 50 per cent. Some of these countries have miniscule areas irrigated (e.g. Gabon or Central African Republic), but others are among the countries with relatively large irrigated areas. In both Madagascar and Nigeria (after Sudan the two countries with the largest

## Irrigated Areas in Sub-Saharan Africa, 1982

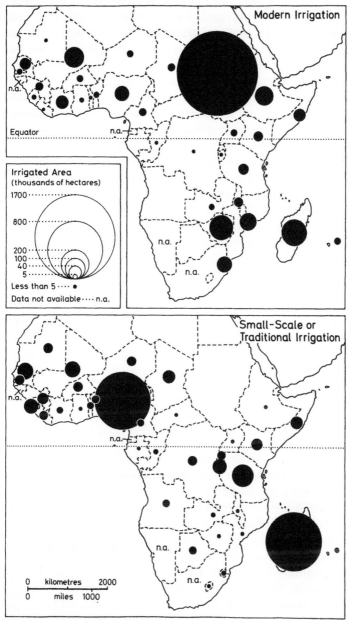

*Source* :FAO Investment Centre (1985) *Irrigation in Africa South of the Sahara*, FAO, Rome.

**Figure 4.1** Area of modern and traditional small-scale irrigation in Africa

total area under irrigation) there are 0.8 m ha in the 'small-scale/ traditional' sector (83 per cent and 95 per cent of the totals in each country). Between them Nigeria and Madagascar hold 67 per cent of all 'small-scale/traditional' irrigation in Africa (1.6 m. ha). In order they are followed in importance by Tanzania, Senegal, Mali, Sierra Leone, Sudan, Burundi, Chad, and Somalia. Each of these countries has over 40,000 ha of 'small-scale/traditional' irrigation.

It is important to remember the provenance and probable inaccuracies in these figures, and also the fact that taken as a whole Africa has a remarkably small area of irrigation – compared, for example, with the South Asian countries of India, Pakistan, Bangladesh and Sri Lanka. Nonetheless, they make two things abundantly clear. First, indigenous irrigation is widespread in Africa. Second, it currently has an important position in terms of the overall irrigated sector in Africa. From these follows a third obvious conclusion, that indigenous irrigation has a vital role in the development of water resources in Africa. Happily, this view has received increasingly widespread support from development agencies in recent years. Less happily, as we shall see, past developments have done much to harm indigenous water management by people in Africa's wetlands. However, before going on to discuss the nature and causes of this problem, it is important to gain a better picture of the indigenous irrigation practices themselves. That is the purpose of the rest of this chapter.

## Farming wetlands

The most basic forms of indigenous irrigation are undoubtedly those involving cropping on rising and falling floods in wetland environments: floodplains, swamps and deltas. Flood cropping embraces a number of distinct practices, including farming on the rising flood (*crue* in French), on the falling flood (*décrue* in French) and farming systems which involve defence against salt water in coastal environments (Table 4.1). Cultivation on the rising flood involves planting before the flood arrives (rice often germinating with the arrival of the rains), and harvesting either from canoes or after the flood has fallen. *Décrue* agriculture involves the use of residual soil moisture left by retreating floods. Floodplain soils often contain clay, and retain water well. Furthermore, water is usually left in backswamp areas and pools long after the river level has fallen. These can be enhanced by human-made bunds to retain water. Farmers become adept at judging the likely duration of water and soil moisture in these areas, and plant suitable crops as the water makes this possible. Bunds are also used in estuarine environments such as the Basse Casamance in Senegal, to keep brackish water off rice fields.

Flood cultivation is a high-risk high-return activity. Floodplain wetlands are highly productive in ecological terms compared to the drylands which surround them. Their productivity is partly due to the fact that floodwaters go some way to meeting evaporative demand, and hence allow plant growth for a longer period. Plants in floodplains are therefore able to use more of the available solar energy, and their productivity is higher. Indeed, tropical swamps are among the most productive ecosystems on earth. In addition, floodplain wetlands are fertile. The annual inundation involves the deposition of silt and other solid material carried by rivers, and the dissolved load of the floodwater. The Logone-Chari Rivers lose 20–60 per cent of its suspended load and 15–35 per cent of their dissolved load when they inundates their floodplain in Cameroon. This can allow continuous cropping in such wetland environments, without the fallowing which is so widely necessary in drylands.

The down side of this productivity is the associated risk. Farmers planting before the floods have no way of knowing exactly when the floods will arrive, nor how extensive they will be, how deep and how long they will stay. The year-to-year variability in river flows in Africa is very great, and farmers are exposed to the risk which follows from it. Floodplains are therefore hazardous environments as well as productive ones. Wetland cultivation can also be very demanding of labour. Mende farmers in Sierra Leone interviewed in 1978 agreed that 'swamp work was harder and more uncomfortable and that swamp rice tasted inferior to upland rice'.[13]

Of course, farmers have various tricks with which they play games against nature. First, most floodplain farmers have a range of crops of different flood tolerance, and varieties of the same crop (e.g. rice or sorghum) with different ecological requirements. It is common to find crops with different requirements planted together in the same field as a way to minimize risk. In the Sokoto valley, for example, it is not uncommon to see rice varieties which need a lot of flooding mixed with a kind of red-seeded sorghum which can tolerate fairly prolonged flooding but does best on a little, and millet, which will not stand inundation for more than a day or so. If floods are high, the rice does well, if medium the sorghum does best, if they fail, the millet comes through. Of course, in reality it is more complex than that, but the principle is one the Hausa farmer would readily recognize: mixed planting and knowledge of crop requirements enables them to deal with the risk inherent in unpredictable flooding.

Second, farmers have the benefit of experience of past patterns of flooding. They are likely to have a good idea of which parts of a floodplain are usually flooded, and how long for. While river discharge is highly variable (and, as discussed in Chapter 3, extremely hard for professional hydrologists to predict), floodplain residents have a large stock of observed and learned knowledge to call on. Farmers use this knowledge of probable flooding conditions to help them decide what to plant and where. Of course, this knowledge is far from infallible, and the capacity to respond to variations in flooding is certainly limited. The droughts of the 1970s and 1980s in Africa had a drastic effect on wetland farmers and other producers.

Knowledge of crop ecological requirements and flooding patterns are matched in most cases of floodplain cropping by a detailed appreciation of the variation in land types in the floodplain. Thus the Marba of the middle Logone valley in Cameroon have a complex system of land classification which takes account of both soils and natural vegetation cover.[14] Unflooded land, called *ambassa*, is of two types, unfallowed garden plots (*bugonlan*) and more distant plots which are fallowed. There is then a category of unflooded but seasonally damp land (*temzeina*), and a wide range of flooded land (*fulan*). Nine different land types are recognized within this flooded category. Each has a different name and is recognized by the characteristics of soil colour or texture or vegetation cover. Each land type has a distinct flood regime (with particular attention being paid to the source, timing and duration of flooding) and each has implications for cropping pattern and the need for fallowing. In the Niger Inland Delta, for example, farmers move between fields in response to the depth and duration of flooding of depressions and lakes.

Flood farmers also spread risk by exploiting environments outside as well as inside the wetland. Paul Richards' work on rice farmers in Sierra Leone shows the importance of their ability to move 'up and down slope' to exploit sites with different soil moisture, fertility and drainage characteristics.[15] He sees this movement as part of a 'rolling adjustment'; to nature, utilizing the availability (and knowledge of) crop varieties. The eventual cropping pattern is not the result of premeditated design based on some static set of skills, but 'a historical record of what happenend to a specific piece of land in a specific year'.[16] Short-duration rices are planted on river terraces and lower slope soils which retain water to grow using residual soil moisture and wet season runoff. Medium-duration rices are grown under rainfall alone on well-drained upland soils, and long-duration varieties are grown in valley swamps or water courses about the middle of the rainy season to grow (if they survive) in deeper water. Straddling ecological boundaries, yet knowing something of the risks and opportunities in each, rice farmers innovate and survive with skill in environments both hazardous and potentially productive.

Farmers in floodplains not only tackle risk by applying their knowledge of environment and environmental variability, but also spread their options into different economic activities. In most cases, floodplain farmers are also fishermen, herders or dryland cultivators; sometimes all three. Just as the wetland provides an additional option for dryland farmers in times of drought, dryland agriculture can provide an important fall-back for wetland farmers in times of flood. Indeed, the balance between the two options, and others, can be very variable even within one region or ethnic group. Once again, flexibility is a key ingredient of success in indigenous production systems.

Some of the flood-cropping practices which occur in West Africa appear to be of great age, and are based on crop plants cultivated within Africa. Much of the rice grown in Africa is Asian Rice (*Oryza sativa*), which was probably introduced by the Portuguese from the sixteenth century onwards. From about the 1930s onwards, improved varieties of rice began to be introduced by colonial agricultural departments, for example among the Mende of Sierra Leone.[17] There is also a West African rice (*Oryza glaberrima*) which was probably domesticated some 3,000 years ago in the Niger Inland Delta. The French scholar Roland Portères suggests that there were two 'hearths' of domestication, a main one in the Niger Inland Delta and another in the West, one in Senegambia/Casamance.[18] There may have been a third in the Guinea/Liberia/Sierra Leone Highlands. Archaeological research on the settlement mound of Jenne-Jeno in the Niger Inland Delta has found evidence of rice-growing (in association with fishing and herding) some 2,000 years ago.[19] Excavations on the vast settlement mound of Daima on the clay plains South of Lake Chad identifies sorghum cultivation some 1,300 years ago.[20]

In the Niger Inland Delta (and other areas such as the extensive Logone floodplain) various wild grains occur and are sometimes gathered, for example the wild rice *Oryza barthii*. Studies in the Gourma region of Mali have shown the importance of wild grain crops as food, particularly in damp low-lying alluvial areas, to pastoralists.[21] It is not difficult to imagine plant domestication occurring where such wild species exist and are known and exploited, although it is dangerous to leap to conclusions about how domestication occurred, and by whom and when it was done. Nonetheless, it is clear that agriculture which exploited natural flood regimes certainly occurred in a number of environments some millenia before the present. In both the Inland Delta and the clay plains of Lake Chad broadly similar forms of production to those identified by archaeologists continue to this day. Certainly flood-related rice agriculture was reported by the first European explorers to visit West Africa. The Portuguese explorer A. d'Almada reported rice cultivation on the South bank of the Gambia River (i.e. probably in the Basse Casamance) in 1594,[22] and Richard Jobson recorded transplanting of rice along the River Gambia a few years later.[23]

There is much still to be learned from archaeological excavation and other sources (for example plant genetics) about the history of plant domestication and of cultivation practices. However, there can be little doubt of the antiquity of flood-cropping systems in West Africa. Such floodplains have long supported dense populations, and indeed important centralized states. Jenne and Tombouctou in the Niger bend were important trading cities in the European medieval period. In the fifteenth century the kingdom of Songhai, centred on Gao on the Niger, was the major state of West Africa, while the Fulani state of the early nineteenth century was centred on Sokoto in Nigeria, on the productive floodplain of the Sokoto River. Suitable environments with marked seasonal rhythms of inundation and desiccation occur in fringing floodplains, inland deltas and coastal floodplains. Similar practices exist, on a smaller scale and without the extensive development of either centralized states or dense populations, in some East Africa floodplains.

## Floodplain irrigation

The principles outlined in the last section can be seen working in the hands of farmers in many of the floodplains of African rivers. The fringing floodplain of the River Senegal is among the best studied, with extensive work by a number of francophone scholars.[24] The River Senegal runs in a broad floodplain up to 30 km wide for 600 km downstream of Bakel (see Figure 4.2). Like the rainfall of these regions, river flow, and hence the extent of flooding, are highly variable. The total area of floodplain land is about one million ha, of which the amount actually cultivated in any one year is very variable. In the 1960s (when rainfall was good) about 150–200,000 ha were cultivated. With the drought of the 1970s the area fell to perhaps a tenth of that. The Senegal River rises in the highlands of the Futa Jallon in Guinea, and the flood reaches the lower valley at the end of July. The river rises above bankfull and the water spreads out across the floodplain through August and September. The floods start to fall from October onwards, draining first and fastest upstream, and are gone by the end of December

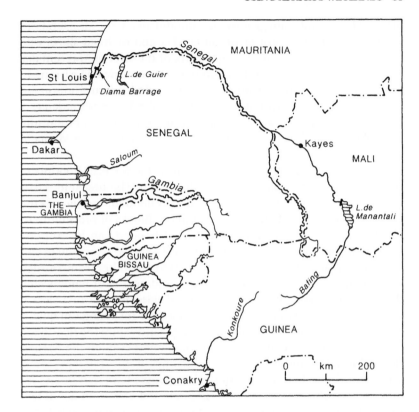

**Figure 4.2** The River Senegal

or January. The floods are thus out of synchrony with the rains (June–September), and the floodplain is cultivated in the dry season using the floodwater and the residual soil moisture in the floodplain soils. The rainfall in and around the floodplain declines downstream from above 1,800 mm in the headwaters to 3–400 mm at the downstream end of the floodplain.

About half a million people practice flood-related cropping in the *waalo* land of the Senegal floodplain. Many of them also keep cattle, and some catch fish. The first and largest group are those who speak Pulaar, the Peul (or Fulani) and Toucouleur. This group makes up over half of the population of the valley, and is diverse. It includes the Peul themselves, who are nomadic cattle-keepers who range tens of kilometres outside the floodplain in the wet season (when grazing is plentiful), but come back to cultivate in the floodplain in the dry season. This group also includes other agro-pastoralists who speak Peul but are sedentary, and people who are primarily (but not solely) catchers of fish. The second main group in the floodplain are the Moors (about a fifth of the total), most of whom are Haratin or freed slaves who established themselves on the right bank of the river (in present-day Mauritania) in the eighteenth and nineteenth centuries. The last main groups are the Soninke, who are sedentary dryland cultivators who reach into the upper end of the valley, and the Wolof who are agro-fishermen who live in small numbers at the downstream end of the floodplain below Dagana, having been forced south by the Moors in the past.

Various classes of land are recognized in the Senegal floodplain, reflecting the geomorphology of the floodplain. The best land is that in the deep clay basins which are most deeply flooded, and stay wet the longest. These are called *hollalde* in Pulaar. Blocks of plots of up to 120 ha are laid out in these areas, and if flooded can be cropped without abandonment for a fallow period. These old oxbow lakes and other floodplain depressions are interspersed with other features, such as old point bars and banks and levees. Land near river banks which floods is called *falo*. This land is typically divided into narrow plots at right angles to the river, their width and length reflecting soil quality (larger on poorer-quality sandy soils). High land within the floodplain is cultivated with rainfed crops. The commonest crop on the deeper-flooded *hollalde* land is a sorghum which matures in 130 days, sometimes mixed with beans. The Toucouleur people have sixty names for different varieties of sorghum. Sorghum is also grown on about half the *waalo* fields, although maize is more common, and a wide range of other crops (e.g. tomatoes, pumpkins and melons, groundnuts, sweet potato and sometimes tobacco) are also grown.

The dependence of cropping on the pattern of floods in the Senegal floodplain is repeated in a number of similar riverine environments elsewhere in West Africa. My own work in Nigeria in the late 1970s concerned indigenous floodplain agriculture in the valley of the river Sokoto in Nigeria (see Figure 4.3), and the impacts of dam construction upon it.[25] As in the Senegal, farmers had a shrewd idea of what depth and duration of flooding to expect on different parts of the floodplain, and they recognized a range of crop varieties suited to different conditions. In the Sokoto case the river floods in synchrony with the rains, and cultivation on floodplain land (like other seasonally damp land called *fadama* in Hausa) was closely integrated with dryland farming and all the other economic activities available. All farmers had access to rainfed land, but not all to floodplain land. Those farmers (almost all of whom were Hausa people) grew several named varieties of rice as well as sorghum varieties which were resistant to flooding. The Sokoto floodplain was perhaps different from the Senegal in that these crops were typically followed by a second cycle of cropping in the dry season, often on deeper-flooded land not available in the rains. This *aikin rani* (literally 'dry season work') was widespread and intensive, and often involved the use of wells and sometimes shadoofs to extend the growing season. This kind of intensive irrigation is discussed further below.

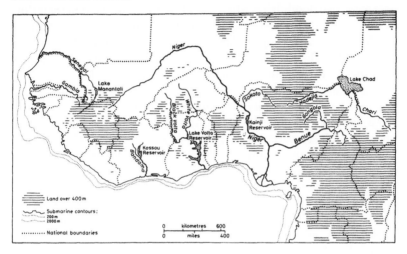

**Figure 4.3** Major West African rivers

Also in Nigeria, both sorghum and rice are grown on seasonally flooded land in the Hadejia-Jama'are floodplain. A number of different rice varieties are recognized in the wetlands, each with specific ecological requirements. Land preparation for rice takes place in the dry season preceding the rains and rice is planted once the rains have begun but before the floods rise. Seeds may be broadcast, or a volunteer crop allowed to grow from seed shed during the previous year's harvest. Once the rice plants have grown to about 12 cm, they can then survive flooding and grow rapidly up to a total height of over one metre. It is essential that the rains precede the flood. Early flooding causes a significant risk of crop failure. Thus the rice harvest is affected by changes in flood extent, timing and duration in the wetlands. Farmers respond to the uncertainty of flooding by constructing defensive bunds (*jinga* in Hausa). Land exposed by falling floodwater at the end of the rains is planted with a flood recession crop which then grows on residual soil moisture. Cowpeas in particular seem to be able to grow in cracking clays long after they appear dry. Cotton or cassava are grown on slightly higher land.

Further east, there are a range of flood-related cropping practices in the Logone-Chari system.[26] The floodplains of the Logone River in

Cameroon are extensive, some 200 km long and up to 40 km wide. The Chari-Logone river system drains the savanna uplands of the Central African Republic, and provides the vast majority of the inflow to Lake Chad. In years of high flow, some water spills Westwards into the Benue system above Bongor, which flows Westwards into Nigeria and the lower Niger River. Downstream of Bongor the Logone and Chari inundate extensive floodplains, where they lose 15–40 per cent of their flow by evaporation, plus 20–60 per cent of their sediment load.[27] Much of the floodplain of the Logone lies on the West bank, and is drained into Lake Chad via the El Obeid River. There is also flooding from channels entering the floodplain carrying runoff from local rainfall.

The detailed system of land classification of the Marba of the middle Logone has been described above. Further downstream, the Mousgoum and Kotoko live on mounds in the floodplain of the Logone. They use various wild grasses (e.g. the wild rice *Oryza barthii*) and catch fish, but also grow bullrush millet on dry ground and sorghum on the rising flood. This grows with the rising floodwater and is harvested by canoe. These practices have declined in recent years, partly because of changes in the flooding pattern of the Logone caused by the polders built for irrigation development by the Société d'Expansion et de Modernisation de la Riziculture de Yagoua (SEMRY).

Around Lake Chad, the black cotton soils of *firki* land have long been used for the cultivation of sorghum (*masakwa*). The soil is impermeable and small bunds (up to 40 cm high) are built at the end of the dry season to retain runoff. Sorghum seedlings are planted out from nursery beds in October, and harvested in the following March. Rainfed sorghum is also planted in the wet season.[28] The decline in the level of Lake Chad in the mid-1980s due to drought which caused the Lake to shrink led to extensive development of wet season sorghum planting on the floor of the lake. Ironically, some of this was carried out by Nigerian government employees from the South Chad Irrigation Project which was itself stranded and dried out by the low lake level. Research by Are Kolawole suggests that the extent of *masakwa* cultivation fell in the 1980s because of land expropriation by large-scale irrigation projects and low rainfall.[29]

Along the lower Waanje River in Sierra Leone, extensive seasonally-flooded riverine grasslands (called *bati* in Mende) are cultivated with rice.[30] There are two seasonal rice crops, a rainfed rice (*sokongoe*) planted on sandy islands away from the river and a floating rice (*kogbati*), which is planted as the floods rise and harvested as they fall. Some of the deep water varieties now planted are higher-yielding Asian rices introduced by the Department of Agriculture from the 1930s onwards. There are also two dry season rice crops. *Bongoe* is transplanted out as the flood falls. *Gbali* is broadcast onto low-lying land to grow using residual soil moisture, and harvested before the river rises. No crop gives great yields, the area suitable for the rainfed crop is small, the transplanted rice makes heavy labour demands, and the residual soil moisture rice is vulnerable to pest and bird damage. Nonetheless, the system is remarkable in providing rice throughout the year.

Flood-related cropping in fringing floodplains is not restricted to West Africa, although it is true that it is most extensive, and best studied, there. It has been recorded in the Omo Valley in Ethiopia, the Tana River Valley in Kenya, the Rufiji in Tanzania, the Zambezi and the Lufira in Zaire. Flood-recession agriculture is also practised in the molapos of the Okavango Delta, Botswana, and the floodplain of the Pongolo River in Natal.[31] In few cases have detailed studies been carried out, but the general principles are familiar: integration of wetland and dryland resources, dependence on seasonal flooding regimes and recognition of different land types and associated flood patterns. The Tonga on the banks of the Zambezi, for example, recognize five garden types. Two are only used in the rainy season (*unda* and *temwe*), a third (*kalonga*) lies in damp areas in small stream valleys and two more (*jejele* and *kuti*) involve flood-recession cropping. *Jejele* land on the banks of the river is cultivated as floods decline, and may in parts be double-cropped.[32]

Flood-cropping is also practised away from major rivers and wetlands, in some surprising places. One is Turkana in northern Kenya. The Turkana are semi-nomadic pastoralists, their mobility and mixed herd structure allowing them to use the harsh and arid bush land of Turkana District. The seasonal flows of the major rivers (Turkwel and Kerio) are vital elements in their resource inventory. The floodplain forests which are maintained by floods and groundwater provide seasonal fodder for livestock (especially protein-rich Acacia seed pods), and also allow some cultivation of sorghum in small fields (100-2000m$^2$) by women. Cultivation takes place in various locations, on river floodplains, in small catchments and on the shore of Lake Turkana and the delta of the Kerio river. The sorghum is quick-maturing (65 days), and cultivation begins in April after rain. In the middle Turkwel, the Ngetebok Turkana grow a range of crops in addition, including maize, cassava, sweet potatoes, beans, gourds and pumpkins. In the delta of the Kerio River there is a problem of soil salinity which is dealt with by growing the halophytic plant *Suaeda* with the sorghum.[33]

The largest continuous areas of floodplain irrigation in Africa are the great inland deltas. In the floodplain of the Bar el Jebel in the

Sudan, Dinka, Nuer and Shilluk people plant sorgum and other crops as the floods recede, in association with rainfed crops on higher ground and dry season grazing on flooded *toich* grasslands.[34] There is also small-scale flood recession cropping of grain (*molapo*) in the south-eastern part of the Okavango Delta in Botswana, based on flooding from the Boro River. More important than either of these, however, is the floodplain irrigation by the Marka and Rimaibé rice farmers of the Niger Inland Delta in Mali (see Figure 4.4), studied in particular by the French geographer Jean Gallais.[35]

The Niger Inland Delta lies within the Sahel Zone. The climate of the Delta is dry (200–700 mm rainfall) and climate strongly seasonal. Flows in the River Niger are also seasonal. Discharges vary from 70 cubic metres per second in May to 5,000 cubic metres per second in October. The annual mean is 70 billion cubic metres, almost half of which is lost in evaporation, evapotranspiration and infiltration to groundwater within the Delta. Agriculture is integrated with fishing and herding, and involves both rainfed and flood-related cropping. Soils are poor and rather acidic. In low-lying areas they are hard to work and crack when dry.

Farming systems are closely adapted to environmental conditions. Farmers make use of different land types and change the location of fields frequently. Both sorghum and rice are planted, with bullrush millet on unflooded land. Where wells can reach the water table, rainfed crops may be followed by irrigated crops in small gardens. *Décrue* sorghum is planted on the falling flood in January, often with other crops such as cassava or groundnuts. Sorghum is sometimes transplanted from raised beds so that best use is made by the growing plants of the moisture left in the soil by retreating floodwater. Rice is planted in the rains in July and August. It grows with rising floodwaters and is harvested as the floods recede between December and February. Different varieties of rice are adapted to different soil and flooding conditions, but several may be planted together to minimize the risk from unexpectedly high or low water levels.[36]

Coastal delta environments are also exploited through flood-related cropping. Rice cultivation in reclaimed mangroves of the Guinea-Conakry coast was reported by slave ship captains in the eighteenth century. Indeed, the skills of these and other rice farmers were much in demand on slave plantations on the East coast of North America. The influence of African indigenous knowledge on rice cultivation in the Carolinas was decisive.[37] Rice is today grown in a number of coastal environments in West Africa, particularly by the Diola or Joola of the Basse Casamance of Senegal and in Guinea and Guinea-Bissau.[38] The same principles of response to flooding apply, with the additional

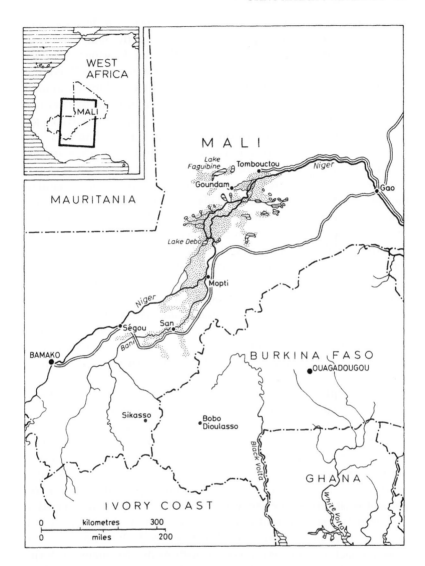

**Figure 4.4** Niger inland delta, Mali

problem of the penetration of saline water. There is considerable cultural diversity among the Joola, depending partly on their contact with surrounding people and their agricultural systems, and also environmental variety within the Basse Casamance. Rice is integrated in different ways into other economic activities such as dryland farming and herding. On valley side slopes different kinds of rice-fields occur depending on the depth of freshwater flooding. Banks and drainage ditches are built to control flood timing, depth and duration. On lower-lying land on the edge of the mangroves, there are bunds to exclude salt water, and canals to supply fresh water. Flowing freshwater from local runoff is used to make the unstable and saline soils cultivable, and bunds to prevent tidal incursion. The extent of penetration of saline water up the estuary of the Casamance is affected by river discharge, and the drought of the 1970s and 1980s has presented a major problem for flood rice production in these coastal environments.

In certain places in Africa indigenous irrigation extends to the exploitation of groundwater. The simplest technology is a well, dug into soft floodplain sediments from which water is drawn with some kind of bucket, typically a half-calabash. This kind of irrigation is extremely laborious, and water can only be lifted about 2–3 metres. It is, however, widely used in Africa. It was, for example, the staple technology in the the Sokoto Valley in Nigeria in the 1970s, where it is presumably of some antiquity. It is also used to irrigate shallots on raised beds in the Volta Delta.[39] Simple wells and bucket lifts are reported from many different areas, notably the dambos of the communal areas in Zimbabwe. These are small damp grassy valleys which occur widely in semi-arid areas of southern Africa (they are also called *vlei* and *mbuga*). They are used for many economic activities, and have become important sources of food through very small-scale irrigation of vegetable gardens.[40]

A more sophisticated version of this simple well technology uses a shadoof, a balanced beam on a trestle, with a weight to counter-balance a bucket. This is more efficient in energetic terms, but is still limited in the depth from which water can be lifted. A man can lift water up to 3 m at a rate up to 4,500 litres per hour, but it is hard labour. The use of shadoofs seems to have been restricted to those areas of Africa which came under most intensive arabic influence, particularly semi-arid parts of West Africa. They are widespread in suitable environments in northern Nigeria and southern Niger, parts of northern Cameroon, and elsewhere in the Benue and Niger basins.[41] These areas were in close contact with the Maghreb and with Arabia for many centuries before north European explorers arrived on the scene, and the shadoof may well have been borrowed through such arabic routes. The shadoof dates back to at least the early nineteenth century in northern Nigeria, and is probably far older.[42]

Animal-powered lifting of water is largely restricted in its distribution to the Nile valley. The ox-powered waterwheel, the sankia or *sāqya*, appears to have allowed resettlement of Lower Nubia in the first few centuries AD.[43] The *sāqya* was widespread in Egypt in the nineteenth century, and the *tabūt* (a water wheel) and the *tambūsha* were also used.[44] Until the Aswan Dam forced the removal of Sudanese Nubians to the Khashm el Girba irrigation scheme, their agricultural economy depended on irrigated cropping (using the sankia) and floodplain date palm plantations.[45] Animal power is also used to lift irrigation water in the Air Mountains of central Niger. Edmund Bernus suggests that garden cultivation is several centuries old in this area. Oxen-powered wells were widespread in the middle of the nineteenth century.[46]

In places, human-powered water lifting for irrigation is being rapidly replaced with small motor-powered pumps. This trend is well-advanced in northern Nigeria, where small petrol pumps began to be introduced in the early 1980s. These are compact, portable and relatively cheap to purchase (even without subsidies). Petrol is relatively plentiful and cheap in Nigeria. The pumps use a comparable technology to that of small motorcycles, which have become ubiquitous in rural Nigeria in the last ten years, so it is relatively easy to find people to mend them. The use of pumps has snowballed well beyond the scope of the original introductory project. It seems that given the right technology, and the right economic environment, long-established indigenous wetland production systems can be rapidly modernised – in this instance with minimal input from formal 'development' initiatives. This issue is discussed more fully in Chapter 8.

## Stream diversion

Africa also has indigenous irrigation of a more formal and controlled kind, involving the diversion of water into canals and its distribution to fields. Such irrigation occurs in a series of locations in the East African Rift Valley, from Jebel Marra in Sudan (which is actually at the eastern edge of the Chad drainage basin) and Konso in Ethiopia in the North through Pokot and Marakwet in northern Kenya to the Sonjo in Tanzania in the South. Comparable systems also occur in the Taita-Taveta and Pare Hills in Kenya and Tanzania and among the Chagga of Mount Kilimanjaro (see Figure 4.5). Although there is little if any evidence of historical connections between the people operating these systems they are physically remarkably similar. The same basic approach is used in other places, for example by the Dogon of Mali.[47] In the Ahaggar massif, agriculture using gravity flow of water in excavated tunnels (*foggara*) spread in the nineteenth century.[48] This kind of irrigation is commonly called 'furrow' irrigation, but in fact is very similar to the hill irrigation systems of the Karakorum in Pakistan and the Himalaya in Nepal and Bhutan so that the phrase 'hill furrow irrigation' is probably better. Irrigation of the shallow valley swamps of Rwanda and Burundi should probably be included in the same broad category.[49]

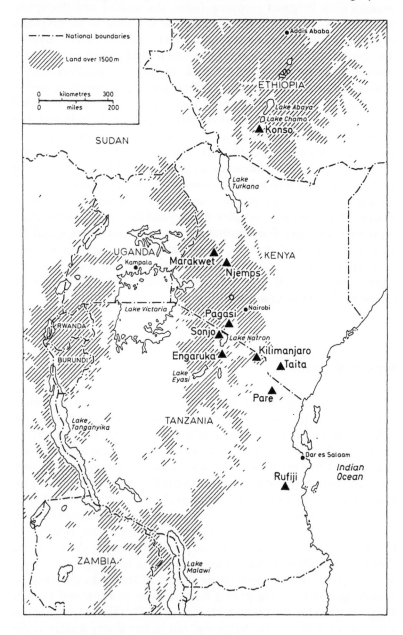

**Figure 4.5** Hill furrow irrigation in East Africa

Like other forms of African indigenous irrigation, there are costs and risks involved in hill furrow irrigation. The construction and maintenance of dams and canal systems are costly in human effort, and make continuing demands in terms of management. In many systems there are problems of water supply, and labour supply, as well as potential problems at least of soil fertility and erosion. In most cases the research has not been done to do more than start to understand the dynamics of these systems, but it is important to recognize the problems and risks with this kind of intensive production.

The existence of this irrigation has not been lost on outside observers. The District Officer in Marakwet District in Kenya wrote of 'the furrow makers of Kenya' in the Geographical Magazine in 1941, and Elspeth Huxley also drew attention to the Marakwet irrigation in her celebration of early colonial years in Kenya.[50] Despite this, the relatively small extent of these systems and the remoteness of the locations in which they occur mean that until very recently they have been unremarked by most engineers and contemporary development experts. Indeed, such people have been known to maintain that they were begun by colonial administrators. Nothing could be further from the truth.

The irrigation of the Marakwet in Kenya is perhaps the most extensive and impressive to a first-time observer. The Marakwet live on the escarpment on the western side of the Rift Valley, overlooking the Kerio Valley some 1,400 m below. They combine agriculture with pastoralism, the women doing the cultivating and the men the herding and the construction, maintenance and operation of the irrigation canals. These take water from small but torrential streams that run down the escarpment, draining the wet highlands of the Cherangani Plateau. Dams made of brushwood, stones and logs are built longitudinally in these streams and water is diverted into canals made of stones and mud. These carry the water along and gradually down the escarpment to the floor of the Kerio Valley where it is used for irrigation, primarily of finger millet and sorghum. The floor of the Kerio Valley is too dry for a guaranteed rainfed crop without irrigation, but there are also gardens on the escarpment itself.

The descent of the escarpment by the canals demands numerous feats of engineering, using log aquaducts to carry water over streams and other channels, or along sheer faces, stone revetted terraces and sluices. Most of the furrows nowadays have sections repaired or rebuilt with cement or other modern material by various agencies over the years, although it is not clear whether these interventions are entirely helpful in the long run. The total length of furrows on the Marakwet escarpment is very large. In one stretch of 40 km of escarpment, there were 40 main furrows totalling 250 km in length.[51] Application of the water to fields is by simple flooding, and it is thought that there is a problem with soil erosion as a result, although this has not been documented. The organization of system management and maintenance seems to be effective, although it may now be under pressure as young men emigrate from the area to work.

Many features of the Marakwet irrigation are repeated elsewhere. On Kilimanjaro, the Chagga irrigate coffee and bananas as well as maizes, pulses and finger millet, although the canals also serve an important purpose in water supply. In Sonjo, streams and springs from an escarpment foot are used for irrigation, which is practised alongside rainfed agriculture and stock-keeping. South of Sonjo is the remarkable archaeological site of Engaruka, which was irrigated in the early iron age. There are stone-terraced village sites at the base of the escarpment, stone-bordered and stone-lined canals through them over an area of at least some 2,000 ha. If the whole area was inhabited at the same time, it may have supported 5,000 people. It was abandoned in about the seventeenth century. It is not certain who the Engaruka irrigators were, or why the irrigation was abandoned, although the possibility that the supply of water from escarpment streams was reduced seems quite likely.[52]

Further North, the Konso live in small towns of 1,000–1,500 people in a highland area of southern Ethiopia. Annual rainfall is low (660 mm), falling in two periods from February to June and from September to December. Fields are small and terraced with stone. Terraces can be 1–1.5 m high and 2.5–3 m wide. Millet, maize, bananas, papaya and other crops are grown. Some fields on hillsides are rainfed, others are irrigated by channels between 100 m and 1 km long. Fields along river channels can be cultivated permanently. Fields are fertilized with animal and human manure, ash and household refuse. Cereals and leguminous crops are intercropped. Storm flows are diverted onto fields, and C.R. Hallpike provides a dramatic description of the intensity with which such water is used: 'even in the night, if there is a thunderstorm, they turn out in the dark and run naked through the pouring rain to their fields to see that the water is flowing well over the soil, and not running to waste'.[53]

## Fishing and grazing in floodplain wetlands

As we have seen, many farmers in African wetlands take part in other activities, thus broadening their range of options and sources of subsistence and income. Of these activities, one of the most important is fishing. Indeed, fish production is a basic element in the economy of many African wetlands, particularly inland lakes and various kinds of floodplains. The life-cycle of many fish species is linked to seasonal flood regimes. The inundated floodplain is rich in nutrients, and local runoff sweeps further nutrients into the floodwaters. As a result there is rapid growth of aquatic vegetation and a bloom of microorganisms and growth in invertebrate numbers. This provides plentiful food for fish, and in many species reproduction is timed so that they spawn when rivers are in flood. Not only is food varied and abundant, but there is a measure of protection from predators. Young fish grow rapidly. Fish making a movement of this kind into the floodplain are said to undergo lateral migration. They may move upstream a short distance before dispersing into the floodplain, although tagging experiments show that some fish will move substantial distances. For example, fish move up the Logone-Chari system from Lake Chad to spawn.[54]

As the floods recede, both adult and young fish of many species move back to the main river channel, and eventually to standing pools where they survive the dry season. There is significant predation both from fishing people and animal predators (particularly birds and other fish) at this time of year. Some species, such as the lungfish and some catfish, are well-adapted to surviving conditions low in oxygen through the dry season. Not all species breed in one go in the floods. Some species, such as the cichlids which brood their young in their mouth, lay eggs in small batches, and usually start before the floods rise. Others, which feed on zooplankton, (e.g. small clupeids) spawn in pools on the falling flood when zooplankton is most abundant.

The amount of fish in a floodplain will obviously vary very greatly through the season. It will be lowest at low water. Studies in the Kafue Flats in Zambia suggest that the biomass of fish falls to about 60 per cent of the wet season peak in the dry season (57,000 tonnes as opposed to 96,000 tonnes).[55] Estimates from African floodplains suggest that fish biomass can be very high, for example 44 kg km$^{-2}$ on the Kafue, up to 220 kg km$^{-2}$ on the Chari in Cameroon , and 20–70 kg km$^{-2}$ in the Okavango Delta in Botswana.[56] It is clear that much remains to be learned about fish biology and ecology, and about floodplain fisheries. Nonetheless, two things are clear: first, the high productivity of fisheries in floodplain wetlands, and second, the close links between that productivity and the flood regime.

Floodplain fisheries are closely adapted to the seasonal rhythm of flood and fish breeding. The floodplain of the Hadejia-Jama'are river system in Nigeria is small (see Figure 4.6), producing some 6,000 tonnes of fish annually, but it illustrates well the patterns of fishing activity and the linkages between ecology and economy which occur in many parts of Africa.[57] A variety of fishing techniques are used. They include lines with baited hooks, lines with foul hooks, gill nets, clap nets and cast nets. Cast nets are used in open water, usually from a canoe. Until 1980, clap nets were made of string, with a mesh size of about 2.5 cm. Now much finer nylon nets are used, which catch much smaller (and hence younger) fish. A variety of fish traps are used in this area, often in association with specially-built bunds. Brushwood bunds occur every 1–2 km on most channels within the floodplain, particularly where water drains out of a lake into a channel, containing basketwork traps (from 30 cm up to 1 m broad and 1.5 m long). In places earth bunds are built over a metre high and up to 400 m long to retain water (and fish) as floodwaters recede.

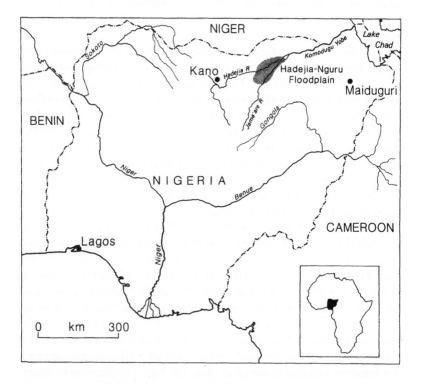

**Figure 4.6** Hadejia-Jama'are floodplain, Nigeria

The fishing gear and techniques used depend on the environment (shallow or deep pools or, channels), and changes through the year. Gill nets for example are used when water is deep in the wet season, and clap nets in the dry season. The best fish catches with nets are in November and December, as the flood falls. At this time a single fisherman can land 15 kg of fish per day. Net catches fall to a maximum of around 6–7 kg per day in the mid-dry season and to 2–3 kg per day at the end of the dry season. Catches in traps peak during the lateral migration of fish which takes place over 4–5 days on the falling flood. In one village 30 traps are operated, and between 54 and 67.5 metric tonnes of fish can be caught over that 4–5 day period. Most of these are juvenile fish. Even quite large catches of such juvenile fish can have a relatively small impact on overall population, compared to catches of adult fish before the breeding season.

Some African wetlands support large fishing communities. The FAO estimate that there are over 60,000 fishermen on the Niger River (50,000 fishermen in the Niger Inland Delta, plus another 6,000 on the River Niger in Niger and Nigeria, and 5,000 on the Benue), and that together they produce 120,000 tonnes of fish per year, of which 75 per cent comes from the Niger Inland Delta. Present patterns of fishing have in many cases been subject to considerable change, both in terms of catch effort and location. On the Logone-Chari system catches rose from 30–40,000 tonnes per year in 1962 to 75–80,000 tonnes in 1970 as fishermen moved northwards into the delta of the river. Fishing effort rose by a factor of thirty. The fish catch peaked in 1974 at 220,000 tonnes, and subsequently fell by half.

Fishermen also come to the Hadejia-Jama'are floodplain from elsewhere. For example one village was formed of 40–50 fishermen, most of whom moved from Sokoto State in northwest Nigeria in about 1953. Fishermen from Hadejia themselves travel in the dry season to fish elsewhere, for example to Lake Chad. Other fishermen make return visits to the Hadejia-Jama'are floodplain, often taking part in fishing festivals at the end of the dry season. Changes in fishing effort over time make it very difficult to assess the status of the fish stocks. Thus, for example, fish catches slumped in the Niger Inland Delta in the 1970s, but it is hard to separate the effects of reduced floods on fish recruitment and those of increased catch effort and possible over-fishing.

Surveys in the Hadejia-Jama'are floodplain show that most fishermen are only part-time, and although there are some villages which specialize in fishing rather than agriculture almost without exception fishermen also farm. It is estimated that about 12 per cent of households have a full-time fisherman in them. In a further 21 per cent people fish only in the dry season and in 15 per cent people fish only in the wet season. The location of floodplain settlements in relation to land liable to flood suggests that around 25 per cent of all fishermen in the Hadejia-Jama'are floodplain are likely to have access to fishing sites which allow them to catch migrating fish retreating from pools and flooded land during the falling flood. The total annual catch in the Hadejia-Jama'are floodplain is likely to be in the region of 6.0 to 6.5 thousand tonnes and the average fish-catch per day is likely to vary by more than a factor of five at different times of year. The total value of the fish catch in the floodplain is of the order of 45 million Naira per year.

African wetlands also play an important role in sustaining dryland grazing systems. Many wetland people combine agriculture with either herding or fishing, sometimes both. Thus in the Niger Inland Delta in Mali there are about half a million people, a mixture of farmers, fishing people and pastoralists.[58] There are five main groups. In the upstream parts of the Delta there are people who primarily fish. They do grow crops, but they leave people to harvest them as the floods start to fall and they migrate through the Delta to end up in the deep lakes of the northeast (Debo and Walado) at the end of the dry season. In this area are a second group of fishers/farmers who grow deep-water rice in the flood season and move much shorter distances to fish as the floods recede. In addition to this, there are sedentary farmers, and two groups of pastoralists.

The Niger draws water from the Futa Jallon far to the south-west, and the productivity of the Delta as a grazing resource depends on the fact that the period of high flood is different from that of the local rains. The rains fall on the Delta itself between June and September. During this period the floods rise, but they peak after the end of the rains, between October and December. The floods fall through the dry season between January and March, and the Delta is dry between April and June. From December onwards the Delta is extensively used for grazing. Because of the floods, it is able to support over 1 million head of cattle and 2 million sheep and goats, 20 per cent of the total numbers in Mali.

Fulani pastoralists leave the floodplain as the rains begin and graze livestock in savanna grasslands. As grazing resources and water supplies run out, they move back into the delta floodplains and graze livestock on the pastures emerging from the floodwaters. In February/March their stock are grazing the dry lake beds in the North of the Delta, particularly on *bourgou* (*Echinochloa stagnina*), a grass growing in flood-water 1.5–3m deep. This provides an abundant volume of forage: 17–30 tonnes of dry matter per ha , compared to 0.5–3 tonnes in dryland Sahelian pastures. The *bougoutières* were devastated by low floods in the 1970s and 1980s, but prior to that were greatly valued both by pastoralists and by farmers, for whom they provided a barrier against rice-eating fish. There are now projects to replant *bougou* in the Delta.[59]

Use of the Delta grasslands allows a remarkably constant supply of grazing resources. Research on cattle herds elsewhere in Mali and in the Sudan suggests that most calves are conceived in the wet season, and that as a result milk supply is episodic within the year.609 In the Niger Inland Delta, the availability of forage and skillful herd management maintain milk supply through the year.

Similar patterns exist in other wetlands in semi-arid areas. For example, in the Hadejia-Jama'are floodplain in Nigeria, Fulani move into the area in the dry season from dryland pastures both North and South. There is also an important trade in *bourgou* as horse fodder for urban areas in northern Nigeria. In the Sudd, Dinka depend on the pastures of the seasonally-flooded *toich* grasslands. In northern Kenya, Turkana depend on access to dry season graze and browse resources in the woodlands along the seasonal rivers flowing into Lake Turkana. In each case, wetlands sustain nomadic or semi-nomadic pastoralism over a large area by providing key resources at critical times of year.

## The integration of wetland resource use

The uses that people make of floodplains and other wetlands in Africa are diverse and complex. They are also integrated economically and ecologically. One result of this has been the development of specific rules about access to such resources. In the last decade, researchers have become increasingly aware of the importance of common property resources, and the relative sophistication of resource management arrangements in many African environments. There is increasing appreciation that indigenous management systems can work remarkably well. With this, however, has gone the realization that 'development' and the interventions of modern states have in many cases so disturbed existing systems that they no longer work. The degradation that outsiders observe is therefore very often not the result of the failure of existing management systems to control resource use, but their failure to function effectively following changes brought about by colonial and post-colonial governments.

In some wetland areas there is a complex history of the organization, allocation and control of resources. Thus in the Niger Inland Delta, for example, there was a system of resource use control based on fishing and farming territories under the control of founding lineages.[66] These were co-opted and changed when the area was colonized by Fulani people from the twelfth to nineteenth centuries. In the nineteenth century, under a unified Islamic state, there was effectively a regional administrative system managed by heads of Fulani clans. This system effectively provided coordinated management of Delta resources, controlling who could use which resources, and when. It thus set clear access conditions not only for pastoralists but also farmers and fishermen. Pre-existing resource users were tolerated to the extent that they did not interfere with Fulani or resident farming communities of slaves.

However, these nineteenth century rules were not static, but were the product of historical change. They had been superimposed upon earlier systems of resource partition, and were in their turn transformed by warfare in the late nineteenth century and the imposition of French colonial rule in 1893. The French colonial administration ended slavery and introduced a new structure of government within the Delta. It intervened directly in the management of resources in the Delta by controlling the timing of herd movement, and created 'fishing reserves'. For the first time, control of Delta resources was in the hands of people not directly dependent upon them. One effect was to undermine the authority of local resource managers and open up access to resources to new users. Following independence in 1960 these trends increased, with the replacement of customary local government heads with external appointments, the development of cooperatives and the expansion of the resource-control powers of the forestry service originally established in 1930.

The break-up of the Fulani control, the imposition of new rules relating to access to resources defined by modern state bureaucracies and the penetration of the modern cash economy have left a confused pattern of controls over resource use. Outsiders have managed to obtain access to resources, partly because they have better access to central government. The technical services lack the scientific knowledge and the local logistics to manage resources effectively, yet their presence, and that of party and local government, mean that existing systems of resource management have lost authority. There has been a tendency for resources to be 'privatized' and concentrated in the hands of influential groups. The droughts of the 1970s and 1980s restricted flooding in the Delta by up to 75 per cent, putting enormous pressure on the already stretched resources. The result has been episodic violence, for example between local fishermen and outsiders at Lake Debo.

Wetlands and drylands are not isolated in Africa, but integrated. Wetland resources are of great value, both to those who have used them in the past and those who wish to harness them to some new development purpose. Changes in the nature of wetland resources, or in access to them, can create very real hardship and can provoke intense conflict. River and wetland development planning has very often wrought just these kinds of change, sometimes deliberately, sometimes inadvertently. The next chapter looks at that planning process and the ideologies that drive floodplain and water resource development in Africa. The environmental and socio-economic impacts of such development, and the problems of inter-community conflict that can arise as a result, are discussed in Chapter 6.

# CHAPTER FIVE

# Dreams and schemes: planning river development

*'In the meantime, let us approach the problem in the tropics with due humility: there is much to be learned before we can teach' (Dudley Stamp, 1953)*[1]

## The eye of the beholder

In the early 1980s I worked for a year on a project planning the resettlement of people who were going to be evacuated from a reservoir in northern Nigeria. I went to this job with high ideals. Like other researchers, I had watched others plan projects and had found flaws in what they did, but I had little experience of project planning from the inside. I did, however, have a high regard for the idea of river basin planning. As a geographer, I had been taught that the river basin was the natural unit for water resources planning, and that only if the whole stream and river system was planned as a unit was it possible to achieve the best development of the water resources of an area. I believed that river basin planning was the only sensible basis for resource development planning. I also believed, without having thought about it terribly carefully, that planners could and did make useful and effective (and fair) plans about the areas under their control. In my naïvety I had but a limited perception of the disturbing learning process ahead.

River basin planning has been part of the model of water resource development exported by international 'experts' to the Third World, and not least to Africa. It has been widely applied both within individual countries and also as a way of managing the waters of international drainage basins, for example the Mekong Basin in East Asia and the Senegal and Chad Basins in Africa. As such, river basin planning has been very influential in shaping the landscapes of the rural Third World, and the nature of the bureaucracies set up to organize and administer their people. However, it is just a part of a much larger set of ideas about the the development of land and water resources. These ideas are diverse, but form a cohesive whole, an ideology of development, that has had a major influence on patterns of Third World development.

The dominant ideology of resource development gradually emerged through the period of colonial rule in Africa. It has become institutionalized in the last twenty-five years in the work of aid agencies and professional development consultants, bringing ideas drawn from the experience of industrialized countries and using them as a blueprint for African development. As Marston Bates drily noted in 1953, 'the white man's burden in the tropics is not the burden of educating, improving or governing the poor benighted natives; it is the burden of his own culture which he has carried into an alien environment'.[2]

River basin planning, and the ideology on which it is based, reflects the confidence and vision of that culture. It conceives of development on a large scale. Rivers need to be controlled by dams and barrages; agriculture needs to be intensified, development needs to be planned and people to be organized. Marston Bates goes on 'I am always bothered by the Western arrogance, by its assurance that it knows all of the answers and can quite readily fix everything so that the tropical peoples can live happily ever after, if only they will listen. This philosophy underlies all the various programs of international technical assistance that are so popular these days, and especially the programs of the United States which are aimed at the uplift of practically everyone else'.[3]

Colonial experience varied, but in British colonial territories the Second World War created a major watershed in thinking and action in development. Following the passing of the Colonial Development and Welfare Acts in 1940 and 1945 there was an intensification of government activity in British territories in Africa. For the first time in the late 1940s, there were people and money available for significant action by the colonial state. This sudden 'access of official energy' has been described by historians Anthony Low and John Lonsdale as a 'second colonial occupation'.[4] Colonial states had exerted major influences on local people and economies before, notably through taxation and the promotion of cash crop production, but now they were intervening in a detailed and technical way, 'in the name of the efficiency that would raise African production, protect African land from erosion or Africans themselves from disease, and in search of the democratic supports without which no project, however expert, could be implemented'.[5]

The new developmental drive depended strongly on science, both for detail (especially in agricultural research) and in some cases for wider insights into ways to conceive of and organize development. The report by Barton Worthington in 1938, *Science in Africa*, set a new agenda for the involvement of scientists in colonial government, and several colonial powers established new scientific research organizations in Africa in the 1940s, for example the British Colonial Research Council, the French Office de Recherhe Scientifique et Technique d'Outre Mer (ORSTOM), and the Belgian Institut pour la Recherche Scientifique en Afrique Central (IRSAC).[6] In 1943 Culwick suggested that government administration in Tanganyika should be 'merely the mechanism for putting the big scientific plan into action'.[7] By the end of the 1940s all British colonies in Africa were producing development plans.

Among other things, such ideas made 'development' in Africa dependent on outsiders and their specialist knowledge. It became axiomatic to assume that the right place to look for ideas and knowledge needed for the development of Africa was the industrialized world and its science, technology and industry. Africa became exposed both to the limitations of this 'expert' knowledge and to the other attitudes and ideas which accompanied the technical knowledge itself.

One of these extraneous, but extremely influential, attitudes was a dismissive view of the knowledge and skills of indigenous users of African drylands, rivers and wetlands. Of course it is easy in retrospect to caricature colonial views of peasant farmers, and to suggest that all observers saw them as improvident, lazy, ignorant, inward-looking. In fact views were mixed, and there were more admiring voices. The British geographer Dudley Stamp wrote in 1938 that 'a recent tour of Nigeria has convinced the writer that the native farmer has already evolved a scheme of farming which cannot be bettered in principle even if it can be improved in detail and that, as practised in some areas, this scheme affords almost complete protection against soil erosion and loss of fertility'.[8] Others were more cautious, and less generous. In 1946 the French geographer, Pierre Gourou described the rice farming of the Niger valley and the West African coast in his book *Les Pays Tropicaux*. Indeed, he saw the improvement of such cultivation as an important way forward for rural Africa. Yet he had little praise for present practices. Rice farming on the Niger was '*fort primitive*', the work of '*une civilisation en arrière*', and the intensive vegetable gardening around Kano in Nigeria he dismissed as the product of 'a more evolved civilisation enriched by elements borrowed from the Mediterranean'.[9]

By and large, however, new initiatives in Africa ignored indigenous skills, just as they often rode roughshod over culture and livelihood in their attempt to refashion landscape, economy and society. Very often 'traditional society' was seen as something in need of transformation. The 'subsistence economy' was taken as a static base on which to introduce change. Institutions such as cooperatives attempted to re-create a past that was largely mythical.[10] Such ideas persisted after the end of colonial rule. A review of 'Operation Arachide-Mil', a dryland agricultural programme developed in Senegal in the 1960s, concluded 'the decisions of modern political authorities have rarely been concerned with, or capable of, using traditional strategies to their advantage, or exploiting traditional techniques and the knowledge of the environment on which they are based'.[11] The attitudes that gave rise to such critiques can still be found today.

## Grand designs

Another feature of many post-war initiatives, driven by the new developmentalist ideology, was their large scale. This stemmed partly from a self-conscious attempt to apply the forms and scale of military organization to peacetime, and partly to the extension of Fordist ideas about the efficiency of large-scale production in industry. In his book *The Earth Goddess*, G.H. Jones commented as long ago as 1938 that 'the present demand for large-scale farming operations derives its force from an industrial source and involves really a simultaneous alteration of farming methods by the application of machinery to agriculture'.[12] In his account of the Niger Agricultural Project, begun in 1948 in Nigeria, K.D.S. Baldwin spoke of the way in which the war-time results of mass-production methods and large-scale projects had 'impressed the public mind', as had the notion that centralized control of production could be efficient in agriculture as in other sectors of the economy. The Niger Agricultural Project was the result of trying to apply war-time methods to peaceful ends, of 'loose thinking and vague aspirations' and the fallacious notion that 'the bigger the scheme the better the results likely to be obtained'.[13]

The persistent urge to 'modernize' African agriculture dates from this time, paralleling major shifts in the economic and technical organization of agriculture in Europe and North America through mechanisation, use of pesticides and fertilizers, increasing scale of operation and the supply of raw materials to an industrialized food system. The superiority and desirability of such techniques were not questioned. It was simply a matter of making conservative African peasants adopt them.

However, in the main African rural people were reluctant to be co-opted into wholesale modernization. This led to dramatic operations by colonial (and later independent) states to implant islands of modernity in the unresponsive matrix of peasant agriculture. It was a sustained, costly and in some ways visionary effort. Unfortunately, it rarely worked. In the developmentalist ideology Africa was considered a 'virgin land', as was Soviet Central Asia in the 1960s or the Amazon is today by their respective states. Once development schemes were begun, their poor performance was lost in a welter of wishful thinking, part of an emerging national fantasy.[14] The fantasy has persisted, locked into the thinking of many professional development consultants.

Many of the grandiose agricultural projects proposed and developed from the 1940s onwards were failures. The Gambia Egg Scheme, which involved the clearance of 10,000 acres of bush to plant sorghum for poultry feed in the hope of producing 20 million eggs for the British market collapsed following crop failure in 1949 and competition from cheaper South African and European eggs.[15] The Groundnut Scheme in Tanganyika, proposed in 1947, was supposed to involve some 3 million acres of mechanized groundnut production for the British edible oil market. The technical problems were legion (not least the failure to clear stumps and plough using converted Sherman tanks from the war: a real case of swords into ploughshares), and poor soils and poor rainfall (neither predicted because of poor technical appraisal) caused disaster. After repeated attempts to create a rescue plan and bitter debates in Parliament in 1949, the scheme eventually collapsed with huge losses in 1951. According to one recent commentator, more groundnuts had been imported as seed than were ever exported.[16]

The Niger Agricultural Project was another failure, one which its planners (to their credit) allowed to be thoroughly appraised. Like the Tanganyika scheme, it involved rainfed cultivation of groundnuts, in the Middle Belt of Nigeria. The scheme was a joint operation between the Commonwealth Development Corporation and the Nigerian government, 'an attempt to combine the benefits of large-scale mechanized cultivation with peasant farmers occupying separate holdings but working cooperatively under supervision'.[17] Plans were for a massive cultivated area, with a pilot project of 26,000 ha (65,000 acres) in gross area, of which 12,700 ha would be cultivated. In the event there were huge problems of mechanization, particularly with land clearance, and by 1954 only about 4,000 ha had been cleared. Settlers were few and reluctant. By 1951–2 it was already clear that the project would not pay. By 1954 the operating company was in liquidation and the project had folded.

Herbert Frankel commented that the project showed 'the twin dangers in all development projects of grandiosity and arrogance'.[18] The first danger 'shows itself in the facile belief that an ignorance of the facts of an economic situation can best be met by throwing into the fray ever more machines and men to rout them'. The second is due 'to the propensity to regard as irrational those actions of others which we do not readily comprehend'. In fact, development is a process of experimentation wherein projects should 'evolve at a pace suited to human and environmental circumstances'. Those farmers who refused to be drawn into the project 'were neither fools nor knaves and they were certainly not acting irrationally'. As Frankel points out, 'when all is said and done the often despised African peasant comes out as by no means the least resourceful and reasonable in this story of conflicting aims and muddled thinking'.[19]

However, although projects failed, the ideology of large scale development survived. The 'conservative peasant' was blamed for many failures as was the African environment and (more recently) the catch-all scapegoat of corruption. All these, of course, play their part, but the planning process itself is also important. Even now it is rarely scrutinized closely by those involved. The failure of a project does not by any means always provide the springboard for re-appraisal and new ideas. Commenting on the failure over several centuries in attempts to grow cotton in plantations in Sierra Leone, Christopher Fyfe comments: 'it appears than once experts' insights are firmly entrenched they need no aid from empirical evidence: indeed they can light-heartedly ignore empirical evidence entirely'.[20] Projects gain a life of their own. No development professional wants to admit to having planned, built or run a bad project. No administrator or politician wants to admit that money already spent has been wasted. As a result the lessons that might be learned from failure are often buried. There is still little opportunity for either individuals or institutions to learn from past mistakes, despite the abundance of material. Agencies like the World Bank appraise their projects, and these appraisals sometimes contain damning indictments of poor planning and development. However, such information is (perhaps unsurprisingly) regarded as highly confidential.

The importance of learning from experience in project design is recognized. J.R. Rydzewski argues that 'the planner-designer gains expertise in his profession not only by producing over his professional life more and more project proposals, but by observing how his proposals have turned out in practice'.[21] Nonetheless, there is little doubt that the professions most closely involved in development project design and construction have not in the past placed enough emphasis on the need to discuss past projects in the spirit of open learning from mistakes. Most design and construction work is done by private companies, and admitting (and sometimes even analysing) mistakes runs contrary to the dictates of commercial practice. The individuals who work on development schemes often move on before projects are commissioned and have no chance to appraise their own work and learn from it. It is not therefore very surprising if they tend to replicate the same ideas in other places with little or no experience of how they work. The ideology of resource development is powerful on prescription, but weak on learning. It is no wonder that the projects it produces are so frequently unsuccessful.

## Water resource development

Nowhere was the colonial failure to appreciate indigenous knowledge so great, and the enthusiasm for large-scale projects of a novel kind so frequently unsuccessful, than in the field of water resource development. Much of the lack of success of such projects can be attributed firstly to a lack of understanding of environment or existing agricultural practices and secondly to poor technical appraisal. To these might be added a third: a sublime confidence bordering on arrogance.

There are many instances of similar lack of awareness of the past on the part of colonial administrators, and their successors. One case in point concerns the Il Chamus in northern Kenya, studied by David Anderson.[22] The Il Chamus are Maasai-speaking people living in the area of Lake Baringo in the Kenyan Rift. This area is dry, reasonable rangeland but marginal for agriculture. In common with the rest of the arid and semi-arid lands in Kenya Baringo is regarded as of low productive potential, with poor prospects for rainfed agriculture. However, to ivory caravans in the nineteenth century the area was known as a place to stock up with grain. This was grown by irrigation on the flats at the south end of Lake Baringo. Water was diverted from the Perkerra River into a complex channel system. In the nineteenth century, Il Chamus absorbed Maasai pastoralists in times of crisis. However, irrigation was viewed as a temporary expedient and through the century there was a considerable turnover in labour for irrigation.

From the 1880s Il Chamus irrigation began to decline, as the ivory trade shifted North and British rule allowed Il Chamus grazing to expand. The final blow to those still irrigating fell in the first decades of this century when the irrigation works were devastated by flooding in the Perkerra River in 1917. Irrigation was abandoned, although there were small-scale attempts to irrigate again, for example in the drought of 1943 and 1944. However, Il Chamus irrigation had planted a seed in the official mind. Concern about over-stocking and food shortages in the area led to a government proposal for a large-scale irrigation scheme on the Perkerra River in 1936. In the event this proposal did not win the support of the Colonial Office in London, but as one account puts it, 'from this time onwards, it is as though the idea of irrigation from the Perkerra river had a life of its own'.[23]

The idea was resuscitated in the 1940s as part of the new wave of development planning, and a pilot irrigation scheme was eventually built in the early 1950s using as labour people detained under the Mau Mau Emergency. However, the government did not involve the Il Chamus, but brought in people from elsewhere as tenants, as Anderson describes, 'the traditions of irrigation at Lake Baringo were swept aside by modern technology and the desire to maximize productivity and increase economic profitability'.[24] Only after 1956 were Il Chamus and Tugen accepted as tenants. The Perkerra scheme itself has been unsuccessful economically. By 1968 the capital cost stood at £0.27m, recurrent costs at £0.33m and revenue at only £0.095m. The scheme still exists, still a financial burden to the Kenyan state. It has suffered extensive technical problems, especially unreliable water supplies due to the diversion weir, rising salinity and poor crop performance. There have been many desertions by tenants.[25]

Development projects proposed by outsiders are often attempts to replicate the landscapes, economies and technologies experienced elsewhere. Often they are unsuccessful. Research by Paul Richards on colonial policy on rice in Sierra Leone provides ample evidence of the power and persistence of such preconceptions.[26] Rice was Sierra Leone's staple crop, and most of it came from upland shifting cultivation farms. Early rains and a poor burn of upland rice fields, followed by the outbreak of a 'flu pandemic just before harvest, led to a disastrous rice harvest in Sierra Leone in 1918. There were severe shortages of rice by the following July, and rioting in Freetown. The Governor, influenced by his previous experience in Malaya, proposed a major shift in focus away from the uplands towards wetlands. He believed that semi-permanent wet rice cultivation with irrigation would not only be more productive, but also (because it allowed a sedentary population) represented a more advanced stage of evolution towards civilization.

With the backing of the Executive Council the Governor sought permission from the Secretary of State for the Colonies to obtain the services of an expert in irrigation, and 'to speed up the "inevitable" evolutionary transition from dry-rice to wet-rice cultivation'.[27] In the event, the Secretary of State for the Colonies took advice from the retired Chief Engineer and Director of Agriculture in Madras. Both thought that a full-blown irrigation scheme was premature without more hydrological and topographic data, but at their suggestion an agricultural instructor and assistant were sent to the Scarcies Delta in Sierra Leone from the Madras Department of Agriculture. Probably to his own and every one else's surprise, he was impressed by what he found. Yields of 4,000 lb an acre were being obtained through continuous cropping without tilth, manure, weeding or irrigation. His report 'suggested that existing methods would be difficult to better without some years of scientific trials'.[28]

The notion that primitive African agriculture could be transformed using Asian technology therefore got short shrift at this time. Instead, the suggestion was made that wetland rice farmers from Sierra Leone itself should be recruited to teach inland farmers about wetland rice farming. During the 1920s this was done, but at the same time new research began to reveal some of the constraints on inland upland rice farming, notably that of labour shortage. Through the later 1920s and 1930s there was increasing respect for the skills and methods of local farmers.

Despite this however, the notion of a major transformation of rice production did not die, but merely slept. It was awakened in 1941 as the Department of Agriculture embarked on a programme to build large-scale polders to grow rice in the Scarcies Delta to meet war-time demand, using money from the new Colonial Welfare and Development Fund. It was a miserable failure. The project tried to control salinity by excluding brackish water from tidal swamps in the dry season using large bunds and drains. However, it was not appreciated that the soils became acid when they dried, and that only inundation by seawater and early rainfall could control this. Heavy rain and flooding caused bunds to be cut in 1945, and by 1948 the decision had been taken to stop further work.

Just after the Second World War the West African Rice Mission toured the British West African colonies, and in each they identified areas suitable for large-scale irrigated rice development. One area they targeted was the swamp and tidal rice production of the middle Gambia River. Double-cropping of rice was proposed, using mechanization, and the Wallikunda Irrigation Project was developed in 1949 by the Colonial Development Corporation. It was one of a series of attempts to achieve increased rice production in the Gambia from the 1940s onwards. It was, in the event, 'an expensive disaster . . . plagued by a Chaplinesque nightmare of basic design flaws'.[29] The project ceased to function, but was followed by other attempts to achieve the same ends: first a state rice farm, then successively in the form of missions from Taiwan, the People's Republic of China and the World Bank. Judith Carney and Michael Watts comment that 'virtually all these interventions have been spectacular failures', with low yields, little if any double-cropping and massive farmer debt. However, attempts to develop large-scale rice irrigation persisted, the Jahaly-Pacharr project being developed in 1984 on the same site as the failed Wallikunda Project thirty years earlier.

Interestingly, the same panoply of international stars has waxed and waned over rice farmers in Sierra Leone in recent decades, with similar lack of success. Paul Richards argues that the spectre of technology transfer from Asia has continued to plague Sierra Leone, most notably perhaps in the attempt by the World Bank to introduce a 'green revolution' swamp rice package to upland rice farmers in the 1960s and 1970s. Two World Bank-funded Integrated Agricultural Development Projects (IADPs) were developed in the 1970s, and initially water control in inland valley swamps was a major element in their work. However, numerous technical difficulties emerged, including poor design, layout and construction (especially the location of head bund), rapid draining of the swamp (sometimes caused by over-excavation to a sandy soil), problems of soil quality (including iron toxicity) and, above all, the creation of acute short-term labour shortages. Rates of abandonment of 'developed' swamps are high, and recovery of loans is correspondingly low.[30]

One of the most influential water developments in colonial Africa was the Gezira Scheme built in the Sudan between the two world wars (see Figure 5.1).[31] The notion of large-scale irrigated cotton cultivation in the Sudan was discussed at various points through the nineteenth century, and under pressure from Lancashire cotton millers and the British Cotton Growers Association development at Gezira began with irrigation by pumps before the First World War. Large scale development did not begin until 1925, when the completion of the Sennar Dam on the Blue Nile allowed gravity irrigation. The scheme more or less doubled in area between 1958 and 1962 when the Managil Extension (and the Roseires Dam) were built. The scheme continues today and covers some 800,000 ha. It produces half of Sudan's cotton and 60 per cent of its wheat.[32] Although subsequent critics have pointed to many problems of poor management at Gezira, the scheme was widely regarded by both British and French colonial governments as a success, and a blueprint for large-scale dam and irrigation development elsewhere in Africa. Indeed, two officers of the Nigerian government were sent to the Sudan in 1948 to report on the scheme, and their report had a considerable influence on the plans for the Niger Agricultural Project, described above.[33] The Gezira experience, and what was thought to be its success, continues to act as an inspiration for irrigation schemes in Africa to this day.

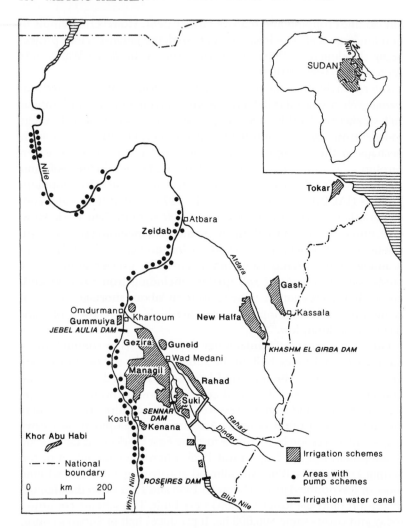

**Figure 5.1**  The Gezira Scheme and other irrigation projects in the Sudan

The attraction of large-scale water resource development was by no means confined to British Africa. Ideas about irrigation in French North Africa (especially Algeria, but also to a lesser extent Tunisia and Morocco) was very much influenced by French colonial experience with irrigation in Indo-China. This experience, refracted through the colonial bureaucracy and the complex ideologies of French colonialism and combined with comparisons with the British experience in India and at Gezira, was important in influencing French thinking about irrigation in tropical Africa. The most notable development that resulted was the establishment of the Office du Niger upstream of the Niger Inland Delta (in what became Mali) in 1932. The idea of large-scale irrigation dates back to a scientific mission to the Delta Intérieure du Niger in 1919. It was suggested that irrigation of cotton by colonists using the waters of the Niger would overcome the problems of drought, and create a self-supporting 'island of prosperity' in the heart of French West Africa.[34] Pilot projects in the 1930s near Bamako suggested that irrigation might be a success, and the Office du Niger was duly established, with a notional target of 1 million ha of irrigation. By 1945 the Markala (Sansanding) Barrage was completed, and 22,000 people were cultivating some 22,000 ha. The same approach was soon applied elsewhere. Shortly afterwards the Richard Toll Barrage was completed on a tributary of the River Senegal, holding back the fresh-water floods between July and November and allowing water to be pumped for irrigation.

In the succeeding forty-five years little of the original vision of the Office du Niger has been realized. By 1962, when the Office du Niger was passed over to the newly-independent Mali, there were only 45,000 ha irrigated, and some 37,000 settlers. These were served by 2,500 full-time employees, and a logistic operation of nightmare proportions, with irrigated land in isolated blocks many hundreds of kilometres from the headquarters. In 1962 the British Geographer Harrison Church wrote that 'many people, including those in the Ministry of Agriculture of the Mali Republic, believe that this money could have been better spent on other projects'.[35] In 1971, cultivation of cotton was abandoned (and replaced by rice), unable to compete in price with dryland cotton grown between Ségou and Sikasso. This has been seen as 'the first blow to the dogma of the productivity of irrigated farming'.[36] Problems have included weed infestation, poor maintenance of structures, over-use of water, and impacts on human health.

The fantastic element within African water resources planning has shown no sign of wavering in the years since the end of colonial rule. Thousands of dams, large and small, have been constructed on African rivers and billions of dollars have been sunk in large-scale irrigation schemes in sub-Saharan Africa. Still the lure of the grand plan attracts dreamers, who conceive dramatic projects on a vast scale to shift water around Africa as if it were a giant playground sandpit. Two current ideas give a sense of the grip that the Grand Plan still has on thinking about African water resource development. They involve dramatic engineering solutions to the persistent aridity of the Sahel, and specifically the low level of Lake Chad, by shifting water North from the Zaire basin.[37]

One such idea, the Zaire-Chad-Niger Scheme, originates within Nigeria, which is at the downstream end of the proposed water delivery system. It is described by one commentator as 'a piece of engineering megalomania'.[38] The scheme was outlined by its originators, J.C. Omulu and V.O. Oke, at a conference on 'water resources needs and planning in drought-prone areas in Khartoum' in December 1986. The basic idea is that a great deal of the 70 billion $m^3$ of water flowing into the Atlantic from the Zaire River is 'wasted'. In the words of the proposers, 'a hydrological linkage of the Zaire, Chad and Niger basins would, therefore, result in a rational redistribution of available water resources and reduction or elimination of wastage of fresh water'.[39] This scheme would involve moving water from the tributaries of the Ubangui River northwards into the headwaters of the Chari system and Lake Chad using a series of dams, tunnels, pipelines and canals, possibly building a dam on the Ubangui to deal with its seasonal flow regime. Further ideas include a canal link to put water into the River Benue for hydro-power generation within Nigeria and a dam in the headwaters of the Zaire and a canal to divert water into the Ubangui, for onward transmission northwards.

The benefits hoped for from the ZCN scheme would include everything from the refilling of Lake Chad (which shrunk drastically through drought in the 1980s), replenishment of groundwater, extensive irrigation, hydro-electric power generation, the 'reclamation of marshes surrounding the Logone River and the lower Zaire', and enhanced fisheries production. To cap this cornucopia of supposed benefits, the ZCN project would 'contribute to the economic and political stability of the subregion by the reduction of enforced migration from drought-prone areas'.[40] The environmental implications of such a project, or any part of it, would be incomparably vast.[41]

The second project makes the ZCN scheme seem modest in its scope. TransAqua is the brainchild of an Italian engineering consultant, and has actually been accepted in principle by the Lake Chad Basin Commission. It basically consists of a 2,400 km canal on the northeast rim of the Zaire basin, intercepting and capturing the flow of its tributaries. These would be carried northwards into the Chad basin, and used extensively for irrigation (no less than 7 m ha) and hydro-electric power (over 30 gwh). The proposed TransAqua canal would be large (carrying 3,200 $m^3$ per second), and navigable. It would therefore act as a North-South transport axis, with freeports along its length providing focal points for new industrialization. It would also intersect with the proposed trans-Africa highway (Lagos to Mombasa). Where they met a 'transport and industrial free-trade zone' would be created.

The proposers of the TransAqua scheme are not shy about the scale of the proposals. The tragedy of the drought 'makes it necessary to think and to act swiftly, guided by courageous policies', and faced with such problems they argue 'we must not be afraid of thinking "big" of "inventing"'.[42] Indeed, 'thinking the future, foreseeing it, is the fundamental task of men of responsibility and goodwill'. TransAqua is portrayed as 'tantamount to a challenge to nature, which has shown itself, and will go on showing itself, in all its harshness and cruelty'. The phrase is well-chosen: the challenge is real.

These schemes are still far from implementation, but they are not by any means beyond the range of the possible. There is considerable momentum, politically and commercially, behind them. The Trans-Aqua scheme has the support of Nigeria and Chad, and the Zairean president has endorsed it publicly. There is at present no information on the vast potential environmental implications of the schemes, either in the receiving or supplying regions, nor on the socio-economic costs and benefits. But money is now being spent to start to assess the size of the planning task. The age of grand designs on Africa's water has not passed away. The lure of the dramatic solution to Africa's problems has lost none of its power. Thayer Scudder comments, 'In African countries, river basin development projects continue not only to be the largest projects within specific national development plans but also to receive the direct interest and backing of the heads of state, whose goals for such projects are as much political as economic'.[43]

## River basin planning

The earliest and most extensive transformations of river flows in African rivers, as well as the first systematic attempts to understand and make plans about them, were on the Nile.[44] Engineering works involving new canals and barrages were built in the delta area early in the nineteenth century to allow perennial irrigated cropping, and between 1843 and 1861 two major barrages were built north of Cairo. These works were improved and extended in the last two decades of the century following British occupation of Egypt in 1882. The first Aswan Dam was built in 1902 , and heightened in 1912 to double its storage. On the upper Nile, the importance of the annual flood to the prosperity of Egypt was a major factor in the imperial ambitions of the major European powers in the second half of the nineteenth century. Once Kitchener's army had seized the Sudan and the Anglo-Egyptian condominium was established in 1899, the Under-Secretary of State for Public Works, Sir William Garstin, was dispatched upriver.

Garstin published technical studies of the Nile in 1904. Ideas included dams and water storage on the Blue Nile, on the River Atbara and in the Great Lakes. They also embraced the notion of building a canal to bypass the swamps of the Sudd. Through the present century such ideas have been studied and debated on many occasions. Firm proposals for various engineering works were made in the 1920s, the first of which was the dam at Sennar on the Blue Nile built in 1925. A second dam on the Blue Nile at Jebel Aulia was built a few years later, in 1937, and the Aswan Dam was heightened again in 1934. Meanwhile, Egypt and Sudan had concluded the Nile Waters Agreement which allocated 94 per cent of the available flows (48 billion out of an estimated 51 billion m$^3$) to Egypt.

From 1931 a series of volumes, *The Nile Basin*, containing ideas and proposals for the development of the Nile, were published by the Egyptian Ministry of Works. Volume Seven (1947) proposed the Equatorial Nile Project, involving storage in the Great Lakes and a canal around the Sudd. This idea was studied by the Jonglei Investigation Team between 1946 and 1954, but the idea subsequently lapsed for over twenty years.[45] Egyptian attention switched to the Aswan High Dam (completed finally in 1971), while Sudan concentrated on the Roseires Dam on the Blue Nile (built 1966) and the Khashm el Girba dam on the Atbara (1965). There was also a new regime of water allocation. A Second Nile Waters Agreement was concluded in 1959, granting a larger share to the Sudan. A new report on the Nile published in 1958, using computer analysis for the first time, gave a better picture of the amount of water available. Egypt secured rights to 55.5 billion m$^3$ (an increase of 7.5 billion) and Sudan to 18.5 billion m$^3$ (a rise of 14.5 billion m$^3$ over its previous paltry 3 billion). The Jonglei Canal was eventually begun in 1976, but all work was stopped in 1983. It remains half built. It is visible from space, a long line scored across the face of Africa, a monument to the scope and scale of engineering planning.

The management of the Nile is unique in Africa for its long history, great technical complexity and huge political significance. It has stood as an example to engineers and planners in the rest of Africa of what can be done by bold and technically-sophisticated planning. Elsewhere in Africa, the grand scale of thinking about the waters of the Nile have been extended by ideas about integrated river basin planning from a rather different source.

The important thing about river basins is that they form what is often called 'a geographical unit'. In other words, the river basin encloses the area drained by all the streams and channels which feed a river at a particular point. All the rain or other precipitation water which falls on the slopes of one basin will either evaporate again, sink deep into the ground, or else end up in the river. Drainage basin hydrologists thrive on the measurement of exactly what goes where through the byzantine complexity of storages and slow flows of water (interception storage, leaf drip, stemflow, throughflow to name but a few) but the upshot is simple. What goes in at the top must come out at the bottom, or at least through some more or less predictable (and measurable) route. It follows that the river basin provides an important way to understand the implications of any particular form of human use of water. Any activity which uses water (like irrigation or drinking water supply), or changes its quality (e.g. pesticide runoff or sewage discharge) or the pattern of its delivery further downstream (e.g. dam construction) has implications for other possible uses downstream.

The logic of river basin planning is simply that it makes sense to coordinate the different uses of water in each river basin so that upstream uses do not interfere with downstream uses, and that (especially where water is scarce) it is used for the best purposes. It is clearly ridiculous to dump toxic mine tailings or pulp mill runoff into the top end of a river basin if there are people downstream trying to drink the water, unless there is heavy investment in water treatment. Similarly, it makes little sense to build dams for irrigation if the water is already being used by farmers downstream, or for hydro-electric power generation if by doing so the livelihoods of fish-catchers and farmers downstream are harmed. River basin planning makes a great deal of sense. Unfortunately, as we shall see, reality in Africa rarely matches expectation.

The idea of integrated river basin planning is something that dates back to the work of the Tennessee Valley Authority (TVA) in the American South from the 1930s onwards. However, while the basic principles of thinking about water resources within the context of the hydrological system of a single basin is widely appreciated, there is less understanding of the political and economic context of the TVA. It formed part of Franklin Roosevelt's New Deal, a response by the Federal government to the pain of the Depression, and it involved what was, for the USA, an unprecedented level of 'hands-on' government. The political context was more or less the exact reverse of the 'liberal' policies followed in much of the industrialized North in the 1980s, where depressed regions were exposed to the economic crunch of economic restructuring by policies of 'rolling back the state'. The TVA 'model' assumed a powerful central state closely involved in making investments in natural resource development.

Development was planned and carried out by a centralized authority (the TVA) which was given a considerable measure of independence (particularly from the individual states), and had a wide range of responsibilities from power generation to navigation. The Tennessee Basin covered over 200,000 square kilometres and had a population of 3 million people, over three-quarters of them rural. The TVA developed a range of projects which brought about extensive changes in the use of water and land resources of the basin. In particular, nine high dams were built for multiple purposes (principally HEP, flood control and navigation).

The approach to land and water resources adopted in the establishment of the TVA reflected the power of rational 'resource conservation' thinking as a theme within American political life since the rise of the 'gospel of efficiency' under the previous Roosevelt at the turn of the century.[46] The US Corps of Engineers had first suggested the idea of 'a public and integrated development authority' as long ago as 1914, and the Niagara Frontier Planning Board established in the 1920s provided some sort of model for the TVA. It is this scientifically-based rational approach to resource conservation which has become the basis of the 'model' of integrated river basin planning for which the TVA is renowned. It fitted closely into the scientific-rational view of resource development being forged within at least some of the colonial governments in Africa.

## River basin planning in Africa

The implications of the supposed success of the 'TVA model' were being pointed out as early as 1944,[47] and the model was adopted essentially unchanged by a United Nations Panel of Experts in 1958. However, it is often forgotten that the TVA experience had as much to do with the intervention of a powerful Federal Government in the economic affairs of a poor region as it did with the adoption of the river basin as a unit of planning. Without that commitment, the 'model' is incomplete, and its applicability to Third World countries rather questionable. Interestingly, enthusiasm for river basin planning on the 'TVA model' in the Third World is not based on any long-term analysis of whether the model has actually been successful in the Tennessee Basin case. Certainly, studies show that the TVA developed in ways very different from those originally planned. The original conception was of a broad resource-development authority, but what has been created is a specialized power-generation company. By 1983, 97 per cent of the

TVA's budget was allocated to power generation, and (ironically) much of the power was coming not from HEP but from nuclear power stations.[48] The economic success of the TVA has also been questioned. It can be argued that the TVA was good for the Tennessee Valley at the expense of the wider economy of the American South. Without doubt it became an example of specialized and not particularly integrated planning.

The cautionary tale of actual development in the TVA has done little to dampen enthusiasm for river basin planning as a sound and appropriate model for the Third World. River basin planning has been widely adopted in Africa. River basin authorities of various sorts have accompanied (not always preceded) dam construction and development, for example the Volta River Authority in Ghana and the Akosombo Dam or the Tana and Athi River Basin Development Authority in Kenya and the development of HEP dams in the Tana River headwaters. Implementation of river basin planning began in Nigeria with the establishment of the Niger Delta Development Board in 1960 and the Niger Dams Authority in 1961 (prior to the commencement of construction of the Kainji Dam on the River Niger for HEP in 1964). International river basins have also been developed under the umbrella of river basin authorities, for example the Central African Power Corporation, concerned with development on the Zambezi, particularly the dams at Kariba (between Zimbabwe and Zambia) and Cabora Bassa (Mozambique).

Such international planning of river basins has not proved straightforward. All Africa's largest rivers cross international boundaries. Planning of the development of the waters of the Senegal River basin began in 1938 in the Mission d'Aménagement du Sénégal, long before the countries now concerned got independence from France. There was also a brief Inter-State Committee which involved Guinea in the 1960s, but this broke up. In 1972 it was replaced by the Organisation pour la Mise en Valeur du Fleuve Sénégal (OMVS), set up with much international aid donor support by Mali, Mauritania and Senegal. The Lake Chad Basin Commission has had a less chequered history. It is supposed to provide an international forum for planning development of all the rivers draining into Lake Chad, and involves Chad, Cameroon, Nigeria and Niger. In practice it has met rarely in the last twenty years, and has achieved very little.

Nonetheless, river basin planning is an important part of the framework within which both African governments and First World aid agencies think about water resources. In the last three decades river basin development in Africa has been supported by a remarkably coherent ideology. This has drawn on the changing fashions in thinking about development through the 1950s and 1960s, views of development as transformation and modernization. The ideology of river basin planning has embraced the notions popular in the 1960s of a 'green revolution' which could improve food production and welfare by technical means (based on new high-yielding seeds) without radical restructuring of society. Enthusiasm for river basin planning had its roots in the disciplines of engineering, hydrology, agronomy and economics which have been central to development planning.

River basin development was dramatic and modern, a sophisticated technical solution to the problems of hazardous and unproductive environments, conservative farmers and rural (and national) poverty. Irrigation in South and East Asia, the USA and Mediterranean Europe provided technical training and conceptual confidence to the professional engineers and economists who came to the Third World as consultants for international organizations and companies. It also created products and 'market-place' experience for companies designing and building dams and irrigation or drainage schemes, or selling engineering equipment such as pumps or irrigation sprinklers. The benefits of river basin development were demonstrated to African technicians visiting American or European Universities for postgraduate training or conferences, was promoted by international organizations like the International Commission on Irrigation and Drainage (ICID) and the UN itself, was central to the work and self-image of professional institutions (notably the engineers), and was sold by glossy brochures of international engineering companies. The ideology of river basin planning was coherent and effectively-promoted. It is not surprising that it was adopted so widely.

## River basin planning in Nigeria

The most-developed system of river basin planning in Africa is probably that in Nigeria, although it is one which has (quite rightly) been much criticized. The Federal Government of Nigeria established the first two River Basin Development Authorities (RBDAs) in the arid North of the country (Sokoto-Rima and Lake Chad) in 1973, and followed this with a further seven RBDAs in 1976. There were then sundry boundary changes, and by the end of the 1970s there were eleven RBDAs covering the whole country. In 1984 the military government under General Buhari increased this number to eighteen. Their name was also changed to River Basin and Rural Development Authorities.

When I began work (via a consultancy company) for one of these RBDAs in Nigeria in the early 1980s I knew little of the practical constraints on planning in the Third World, nor about the reality behind the rhetoric of river basin planning. I was soon to find out. My job was to take part in a team planning the resettlement of some 26,000 people about to be flooded out by a dam on the River Gongola, called the Dadin Kowa Dam. The dam was partly for irrigation and partly for HEP generation, and was being built for the Upper Benue River Basin Authority. By the time we started work, the dam was already half-built, and there had been preparatory studies of the problem of the many people who were going to lose homes and land under the reservoir. If the year taught me one thing, it was that the plight of reservoir evacuees represents a particularly acute cost which should be weighed in any cost-benefit analysis of dam construction. That issue, and the problems of resettlement, are discussed in Chapter 6. Here I want to draw attention to some of the background to the task in the form of the bureaucratic and planning nightmare represented by the river basin planning system.

The first shock was that the boundaries of the River Basin Development Authorities (RBDAs) did not relate to actual river basin boundaries. I had expected (reasonably enough I thought) that these boundaries would run along the junctions between different river systems. Not at all; for much of their length they followed State boundaries. Furthermore, for much of the length of the reservoir the state boundary was formed by the bed of the Gongola River. Thus part of the reservoir area was in the area covered by the Upper Benue RBDA (UBRBDA), while the rest fell within the area of the Chad Basin Development Authority (CBDA). This did not matter too much, for the attentions of the CBDA were more than occupied by their attempts to develop irrigation 200 km away on the shores of Lake Chad using water from the Lake. However, the CBDA had to be kept informed of all planned developments, and to be represented on all planning committees.

More importantly, there were two states involved, Bauchi and Borno, both of which would lose land under the reservoir. Each state had its own ministries of water resources and agriculture, and each had to be kept informed and be represented. There was also the need to bring all the Local Government Authorities into the planning process. Furthermore, the headquarters of the UBRBDA was some 150 km to the South of the dam site, in a third state (Gongola). In an attempt to improve local liaison (particularly as the enormity of the impact of the dam on those to be flooded sank in), UBRBDA chose the only senior member of its staff who came from the local area to act as head of the resettlement project. His was an uneasy job, lobbied by different village and District Heads, squeezed from above by budgetary constraints and irritated by expatriate planners with lunatic notions of 'planning from below'.

The result was a grimly unsatisfactory piece of planning. A preliminary report outlining the basic costs of compensation for flooded lands and houses started to reveal the true cost of the dam. This basic cost was of an order of magnitude more than the total resettlement budget, and took no account of the need to actually take action to rebuild settlements and basic infrastructure (roads, wells, schools, clinics), let alone replace livelihoods. A decision was made by the Authority at an early stage that people should be resettled within their own village areas. While this minimized the traumas of longer-distance movement, it effectively prevented any policy of identifying new farmland to replace that lost. The only economic hope for evacuees was that the seasonal movement of the reservoir boundary (the 'drawdown') would allow opportunities for irrigation, and that the reservoir would support fishing. No money was available to investigate or to develop either activity: indeed, at one particularly stormy meeting, a chapter outlining these 'additional responsibilities' was re-named 'additional opportunities', a highly ironic concept.

There were many meetings of huge and cumbersome committees, a huge amount of commuting and vast bundles of papers and plans. In practice the fact that the dam was already under construction meant that there was no room for any kind of participatory planning beyond the basic human contact between Land Rover-borne expatriate and village head. In the event, old disputes about village boundaries flared up as Village and District Heads competed with each other in lobbying for what they saw as prime sites. A highly technocratic 'planning process' rolled forwards and was presented to the Authority while it became increasingly obvious that there was no money to pay for either legal compensation or resettlement in any conventional sense.

I do not know the ending of this sorry tale. My involvement with the project ended, as indeed did my participation in Third World development consultancy for a good number of years. I have not had the chance to return to Dadin Kowa, but I understand that the dam was held up for most of the 1980s by lack of money for its completion. However, when the reservoir eventually floods there is certain to be confusion and hardship, as well as disputes about compensation. These occurred even at the 130 square kilometre reservoir behind the Kiri Dam further downstream on the Gongola, built to provide irrigation water for the Savanna Sugar Estate. Here, four new consolidated settlements were built and some compensation was paid to the 15,000 evacuees for houses, economic trees and built improvements to land (e.g. shadoofs). Disputes focused on the claim that commitments of compensation were not made in full. As at Dadin Kowa, the minimal legal costs of compensation had been underestimated, and the money necessary to pay even this most basic element of resettlement costs was not available. The problems at Dadin Kowa are likely to be very much greater.

However, the critical point here is not the iniquities and perversities of reservoir resettlement, real enough though they are, but the failure of the river basin planning model in the face of bureaucratic and political reality. One reason for this, in both the cases above, is the inadequate nature of project appraisal. In both cases, dam planning and design were technical procedures limited to the disciplines of engineering and hydrology. A little economics was bolted on, but there was no significant input from land use planning, and certainly no geography, sociology, anthropology or environmental science. This technical limitation was exacerbated by the way that the creation of RBDAs was fitted into the project appraisal process. Despite appearances, the UBRBDA was an afterthought, without any influence on the way projects were planned and developed.

The possibility of building an HEP dam in the gorge at Dadin Kowa was first recognized in engineering studies done on the River Gongola by Dutch consultants in the 1950s. By the mid-1970s a feasibility study had been carried out on the site without any consideration having been given to the river basin as a whole. Furthermore the second dam had already been proposed for irrigation further downstream at Kiri. The UBRBDA was created in 1976, and in the following year the Federal Ministry of Water Resources commissioned a study of the land and water resources of the whole basin. Meanwhile, however, work on both dams went ahead. When the river basin plan was completed, it contained severe criticisms of the two dams. It argued that the basin was being developed the wrong way round. Both dams were in the lower part of the basin, but the report said that what was wanted was a dam in the headwaters to control floods. Furthermore, the two dams had been designed independently, and the Kiri Dam could have been lowered significantly (thus saving money and flooding fewer people and less good valley bottom land) because the Dadin Kowa Dam was there. Belatedly, the UBRBDA tried to get the Kiri Dam lowered in 1977, but the contract had already been signed and the design could not be changed.

Both dams were built according to their original designs, quite independent of each other and the broader interests of the river basin. Despite appearances, no integrated river basin planning took place.

Unfortunately, by the time it was built the amount of power generated by the Dadin Kowa Dam was such a trivial proportion of the rapidly rising Nigerian demand that plans to link it to the national grid were scrapped. Sadly, too, as Chapter 6 will discuss, large-scale irrigation of the kind these dams were to serve has proved economically disastrous in Nigeria. An integrated river basin planning approach that took account of available water and electricity resources and demands within the basin would have allowed much more sensitive and effective planning.

This tale of isolated project-by-project planning is not unusual. Similar things took place in other river basins in Nigeria in the 1970s and 1980s. In the Sokoto Basin in the northwest of the country a land and water resources plan was carried out by the Food and Agriculture Organization (FAO) in 1969, long before the Sokoto-Rima Basin Development Authority (SRBDA) was established in 1973. The team preparing this plan was dominated by engineers, hydrologists and soil scientists, and most of its recommendations were focused on the possibility of large-scale irrigation development. However, the plan did take explicit account of the effects the different dams and schemes would have on each other, and it did recommend an overall basin planning authority. Unfortunately, project development went ahead (with tenders produced by a Sokoto Rima Basin Ad-Hoc Development Committee and the Federal Ministry of Agriculture) before that authority was established. The design for the first irrigation scheme (which became called the Bakolori Project) was signed a full year before the SRBDA was established. That design involved accelerating the project construction process and doubling the size of the irrigation scheme. In the process the allowance for downstream rice-farmers was consumed, and the integration of the individual project with the rest of the basin was lost. The construction contract was signed in 1973 (even though the final design report was not complete until the following year), and construction began on the dam and irrigation scheme in 1974.

The implications of this isolation of the Bakolori Project from its river basin context were serious. In particular, there were significant adverse impacts on downstream agricultural and fishing communities. These and similar impacts of dams are discussed in Chapter 7. Again, the critical point is that the structures of river basin planning were not relevant to the process of planning the actual development of the river basin. The whole bureaucratic structure of the RBDA was created after the vital decisions had been made. In this respect the Nigerian RBDAs are not by any means unique. In Kenya, by the time the Tana River Development Authority was created in 1974, plans for the construction of the large-scale irrigation project at Bura in the lower Tana were well advanced. Although the TRDA (from 1981 the Tana and Athi RBDA) became involved in irrigation, the Bura project was run quite independently by the National Irrigation Board. This was not integrated planning. The right hand had no idea what the left hand was doing.

## Ideology and achievement in river basin planning

If RBDAs in Africa have not been effective in achieving integrated river basin planning, or in promoting development that works, it is worth asking why enthusiasm for them remains so strong. There are many reasons, most of them closely linked to one another. First, the river basin planning model is attractive in its own right. In part this stems from the appeal of rationality and planning to the young states of Africa. Most African states adopted a model of a strong central state and strict limited-period development planning through the 1960s and 1970s. Rational resource use based on structured plans was basic to this view of how to organize development.

Linked to this is the influence of the ideology of river basin planning discussed above. In the 1960s, and increasingly in the subsequent two decades, technical decision-makers in African governments shared the attitudes and ideas of their expatriate advisers. They too believed in the need for modernization, and for large-scale high-tech solutions to perceived environmental and socio-economic constraints; they too believed that technical planning skills were sufficient to achieve success in development. Where did they get these ideas? From the Northern Universities where they were trained, from the battery of messages sent out by international organizations, and from the promotional literature of the private companies exporting their talents from the First World. Only now, after about two decades of criticism of these ideas about development in universities and research institutions in the North, are they filtering down through international agencies. Old ideas are still current in many companies and Ministries, and are locked into the memory banks and word processors of too many consultants. Once gripped by the ideology of developmentalism, leopards are reluctant to change their spots.

Jon Moris describes irrigation as a 'privileged solution' in development. By this he means that its benefits were so self-evident to different groups of people that the real costs and benefits of developments were not thoroughly questioned.[49] Irrigation certainly commanded remarkable support from politicians and bureaucrats in Africa and in the international development community, and somehow policy-makers were blinded to the problems and high costs. The same argument could be made for other aspects of river basin development. It was the flavour of the times, and it is only recently that the cost of the policies which resulted have begun to be counted.

A second and related appeal of the river basin planning model is its sheer scope, and the notion that through such institutions a whole range of constraints on development could be tackled together in an integrated manner – and as a result could be solved. If the Nigerian River Basin Development Authorities are again taken as an example, they had an astonishingly wide range of responsibilities. Their functions embraced everything from flood control, irrigation and navigation through to pollution control, fisheries, seed multiplication, food processing and livestock breeding. They could establish grazing reserves, build feeder roads and undertake mechanized land clearance and cultivation in rainfed areas. In fact there was very little in the way of rural development which they could not do.

A third reason for the widespread acceptance of the river basin planning model in Africa is the way in which it has allowed central governments to bypass existing Ministries and planning structures. This might be done in the pursuit of greater efficiency, to bypass corrupt or inefficient bureaucracies, or simply because river basin projects were visible evidence of the power and good intentions of the central state. Economic performance was somewhat secondary. In Kenya the Kerio Valley Development Authority was created in 1979, with wide-ranging powers over existing line Ministries and their responsibilities and powers.

Probably the key attraction of the RBDA model to a country like Nigeria was the possibility that it might enable the central (Federal) state to bypass existing procedures and planning structures and achieve rapid development. This aim owes something to the realization in Nigeria at the time that the state bureaucracies were cumbersome and inefficient. The RBDAs were introduced by a military government for whom the image of technical efficiency of the RBDA model must have held considerable attractions. This appears to be one reason for the change in the title of the Nigerian RBDAs in 1984 to highlight their *rural* development role, presumably intended to be dynamic Federal agencies with broad powers and responsibilities.

In Nigeria this geopolitical role for RBDAs was particularly well developed in the years of civilian rule between 1979 and 1983. The Nigerian RBDAs provided a way in which the Federal Government (and the ruling National Party of Nigeria) could reach over the heads of the State Governments to dispense largesse direct to rural areas. The result was nakedly political decision-making. At one conference, senior managers from the Ogun-Oshun RBDA (OORBDA) called plaintively for protection from 'the wiles and caprices of highly placed political appointees so that projects may be sited where most desirable irrespective of the political inclination of the people in such areas'.[50] Unsurprisingly, perhaps, there were significant problems of overlap and conflict . The plans of the OORBDA and the Oyo State Government were sufficiently disparate that they actually clashed over who should develop the Idere Gorge Dam site. One commentator wrote, 'the conclusion which one draws is that whilst river basin planning is certainly a very valid way of looking at resources such as water, in the process of plan execution, the interplay of power groups can create distortions which become counter-productive to the effective use of these resources'.[51]

In Nigeria the RBDAs also played a particular geopolitical role in providing a way of investing the rapidly rising revenues from oil exports through the 1970s. Moreover, they did so in a way which used 'southern' oil money to create potentially wealth-creating infrastructure in the north. In the years after the end of the Biafran Civil War, that must have seemed a sound idea both to those who believed in building national reconciliation, and those who felt that southern wealth should be shipped North. African decision-makers looking at a dam or irrigation project were often, reasonably enough, hopeful about the future and the project's possible contribution to it. Even the poor economic performance of existing projects could be set against hopes of future benefits, particularly in oil-rich Nigeria, where the massive capital investment could (until the 1980s) be written off in the continuing slurry of oil revenues. Certainly the Nigerian RBDA projects (principally dams and irrigation schemes) were dramatically successful as sinks of large amounts of Federal capital. The Federal Government allocated 2.1 billion Naira (about 1.7 billion pounds) to the eleven RBDAs between 1979 and 1983. It was unfortunate that most of it sank beyond hope of return.

The political role of river basin planning grades into the complex area of corruption. River basin projects are large and expensive. Very large sums of money are paid over, and commissions of every kind tend to become significant elements within decision-making. A.T. Salau comments that, 'without doubt, river basin development planning has been a "success" more for the local and foreign contractors, bureaucrats and rural élites and incipient capitalist farmers than for the local rural populace'.[52] There is little research on this kind of problem, but nobody with direct experience of major project development is unaware of its prevalence or importance.

In Nigeria and elsewhere, RBDAs have been exposed to the worst excesses of bureaucratic gigantism. They have survived, and proved effective covers for corruption but poor promoters of any form of rational or integrated development. One reason for this is the lack of adequate basic hydrological data. A Nigerian writer commented in 1982 that 'most of the newly-established river basin authorities have not collected enough data to warrant the large scale developments that they envisage'.[53] There is also a lack of trained staff to appraise the reports of foreign consultants. As a result, plans conceived by foreign 'experts' working with inadequate data gleaned during short periods of local fieldwork are approved by organizations without the competence and confidence to criticize them. The result is haphazard and risky development, very far from the 'model' of controlled, efficient and *integrated* river basin development.

What are the implications of the poor performance of river basin planning in Africa? Is the principle of river basin planning at fault? The consensus is still that it is not. The Nigerian geographer A. Faniran argues that failures can be attributed to implementation not the central principles.[54] But what use are theoretical ideas about ways to plan development if they don't work in practice? The answer, unfortunately, is that they are rather dangerous. In the worst case they can lead to environmentally damaging development projects which waste scarce capital and damage established economic activities and the livelihoods of rural people. Salau points out that 'the heavy investment in river basin planning has, to some extent, foreclosed alternative ways of improving rural welfare and productivity',[55] and there is no doubt that this is true. However, in a sense, of course, Faniran is quite right. It is no good dismissing the idea of river basin planning because of a few decades of failure, however costly and however grim for those people affected. The importance of integration between development projects has repeatedly been stressed by both researchers and development practitioners. It was, for example, the chief conclusion of a major study for USAID in 1988, and is a central principle of the work of the Wetlands Programme of IUCN.[56] There is an inescapable logic to the idea of integrated planning, and with water resources this has to be done within the context of the river basin.

It is not the principle of planning water resources in an integrated way that is at fault, but the grandiose scale of the plans which have been put forward. It is the lure of complete control over nature, and over the production process, by a centralized technical planning body which has failed. Understanding the interactions of human uses on each other, and the ways they relate to the natural variations of river flow is vital to any sustainable future in Africa. This is very different from the last few decades of experience with the creation of grand plans and commitment of vast funds in the name of river basin development. Such projects are monuments to the arrogance of their creators, most of whom live far away and unaffected by their failure. This may be good business (in the short term), but it promises a poor future for Africa. There are other ideologies to drive thinking about poverty, environment and development. Aldo Leopold said that 'the first rule of intelligent tinkering is to save all the parts'.[57] River basin planning undertaken in that spirit holds some hope for the future. The structures created so far in Africa simply do not warrant such hope, demonstrating planning failure on a vast scale. The implications of new thinking about development will be discussed in Chapter 8. First, it is necessary to consider the impacts of conventional river basin planning, through the construction of dams and of irrigation schemes.

# CHAPTER SIX

# Binding the rivers

*Engineers these days are apt to claim that they could accomplish miracles – if only the ecologists, economists and politicians would let them (D.J. Allen-Williams, 1975)*[1]

## Losing the flood

In 1978 the flood in the Sokoto River in north-west Nigeria failed. Farmers in the valley depend on the annual flood of the river to irrigate their rice crop, and to leave enough moisture in floodplain soils to allow dry season cultivation. The valley has had a high population density since at least the nineteenth century Islamic Jihad, when Sokoto became the centre of Fulani power. Agriculture in the flood-plain is productive compared to dryland agriculture away from the river, but it is also inherently risky because of the natural variability in the timing and extent of flooding. However, the flood varies within more or less predictable limits, with high flows during the rains in July, August and September. In 1978 there was no water at all in the river between 30th July and 8th September. For the farmers it was a major disaster.

There was no mystery about what had happened, except of course to the farmers. The Sokoto River was to supply water to the irrigation scheme being built at Bakolori, 120 km upstream of Sokoto (see Figure 6.1). In order to store enough water to supply the scheme through the long dry season, a dam was built to hold back floodwaters. It was fin-ished in the dry season of 1977–78, and the gates and outlets were closed. In the 1978 wet season, the water usually flowing in the river was held behind the dam in order to fill the reservoir. This rather drastic move is in fact standard practice among dam engineers. It makes sense to engineers to fill a reservoir, and to bring its water into readiness for use, as rapidly as possible. Unfortunately, it is disastrous for those who depend on the water.

In the case of the Sokoto farmers, the problem of impacts on down-stream users of river water was by no means unexpected. The de-

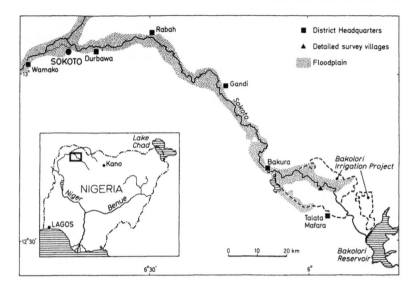

**Figure 6.1** Sokoto Valley, Nigeria

velopment of the project at Bakolori has been described in Chapter 5. Design of the project at Bakolori laid aside earlier plans for downstream releases and an integrated role for the reservoir. The original river basin plan recognized the importance of agriculture in the downstream floodplain, but project design did not. However, as the dam was closed the engineer supervising construction recognized the implications of the fact that river flows were now subject to human control. He warned of the danger to life and limb of sudden releases, and also that stoppage of downstream flows would have serious implications for farmers. Indeed they did.

The farmers themselves took action, and appealed for help through the Sultan of Sokoto. From there the message was filtered through the river basin authority to the engineers and contractors at the dam. As a result, the gates were opened for three days in September 1978 and a token discharge of 23 metres per second was released. After this, flow again fell to zero through the rest of September and October. At least the problem was now recognized, but it was no nearer solution. Some studies were done in floodplain villages in 1979, and the notion of compensation flows was discussed through 1979 and 1980. Indeed my own work in Nigeria was the direct result of the perceived need to understand more about floodplain land-use systems in the valley.

However, there was a lack of information on how much water was required in the form of technical data which engineers could use to calculate flood releases. This was compounded by the lack of sufficient detailed discussions with downstream farmers or surveys of their land to identify flow needs, and the physical complexity of the floodplain and hence of flooding patterns. There were no simple and quick answers to the question of how much water was needed and when. There were also serious contractual problems which prevented effective action. The contractor was unwilling to do anything which slowed down his work (unless paid for it), the river basin authority did not have the money to pay. The supervising engineer had a good grasp of the problems and possible solutions, but unless it risked being made liable for costs could only make recommendations to the authority. These recommendations were not taken up.

Throughout, there was little direct communication between farmers and dam builders. In 1979 a man was deputed to tell farmers downstream 'that they should not rely on a flood flow in the river, and if they still want to pursue the matter it should be done through the Local Government'.[2] Meanwhile, the farmers lacked water. In 1981 they again requested releases, this time through the Governor of Sokoto State. Again the dam gate was opened for three days in September. The maximum flow rate was 500 cubic metres per second, only two-thirds of that thought to be necessary to inundate the floodplain, and this rate of discharge halved over the three days. The effects of these flows were not monitored.

The Bakolori Dam is not unusual in having adverse impacts on people downstream, nor in the fact that these impacts were not predicted or dealt with by the project planning process. The ideology of water resource development described in Chapter 5 has been accompanied in Africa both by inadequate planning and an ignorance, bordering on disregard in some cases, for the interests and needs of floodplain people. However this sweeping assertion needs to be qualified, because it suggests that all impacts of dams are obvious, and all project planners are fools. It must be recognized that the impacts of dams are often complex and difficult to fit into standard approaches to project planning. Understanding the effects of dam construction demands a more complete analysis of floodplain ecosystems and land use, and understanding the failure of project planning to take such impacts into account demands a fairly detailed analysis of the project planning process. This is the aim of the rest of this chapter.

## Dams and reservoirs: gains and losses

The most obvious impact of dam construction is simply the loss of productive land beneath the reservoir, and the costs (economic and human) of resettlement. The area lost beneath reservoirs in Africa obviously varies a great deal, as does its quality. Among the largest reservoirs are the Volta Lake formed behind the Akosombo dam, which covers 8,500 square km, flooding a substantial area of central Ghana (see Figure 6.2). Kainji impounded 1,200 square km, including 15,000 ha of farmland. The Lagdo Dam on the Benue flooded 70,000 ha, including floodplain land stretching 2–5 km on both banks of the river.

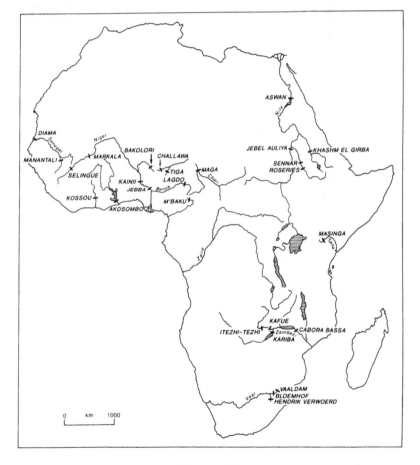

**Figure 6.2** Major dams in sub-Saharan Africa

One measure of the significance of the land loss is to compare it with the area reclaimed, or of improved productivity, allowed by the reservoir. This is often not very impressive. The Bakolori reservoir, for example, covered 12,000 ha. The irrigation scheme for which it stored water covered a net area (total area minus the area covered in roads, service centres, villages, etc.) of 22,000 ha. If we assume that the irrigation area would carry two crops a year (which, it turns out, is optimistic), and further (very conservatively) that only 80 per cent of the reservoir area was cultivated (with 25 per cent of that double-cropped floodplain land), this suggests that 12,000 ha of crops will be lost through the life of the project to create 44,000 ha of irrigated crops. An area equivalent to over 25 per cent of the total area of crops in the irrigation scheme was lost below the reservoir. You do not have to be a trained economist to know that the area flooded by the reservoir must be a significant element in any proper economic evaluation of a reservoir project. Of course, the intensity of land use in the Sokoto Valley, like that of other valleys in West Africa, is much higher than would be typical in east or southern Africa. However, the problem remains. It is remarkable how often this is ignored, or played down, in project planning.

The other major cost of reservoir construction is that of resettlement. The numbers of people affected by reservoirs in Africa is remarkable in contrast to the care taken in developed countries like the UK to avoid flooding houses. The reason for the care is not philanthropy so much as cost. There are clear guidelines for compensation, and built property is valuable. There is also a clear and widely-held view that flooding people's homes in not acceptable, and the political implications of doing so without a very clear and strong reason would be serious. By contrast, development planning tends to run roughshod over the wishes (and low-cost homes) of rural Africans in the name of development and progress. Sometimes very large numbers of people are displaced. The Kossou Dam in the Ivory Coast displaced 85,000 people, the Akosombo Dam 84,000, Kariba 57,000, Kainji 50,000 and the Lagdo Dam on the Benue 35,000 people.[3]

The Aswan High Dam on the Nile displaced an even greater number of people, 120,000 Nubians, both in Egypt and Sudan. The two countries agreed on the construction of the High Dam in November 1959, as part of the negotiations over the Second Nile Waters Agreement. The reservoir eventually stretched 150 km inside Sudan, flooding twenty-seven villages and the town of Wadi Halfa. A total of 50,000 people were evacuated from Sudanese Nubia, and resettled far away on the Khashm el Girba irrigation scheme on the Atbara River. Hassan Dafalla, the chronicler (and, indeed, administrator) of that evacuation, wrote, 'however great the benefits of this agreement, the side effects were disastrous. The Nubians were the only victims – the greatest part of their country being doomed to destruction'.[4]

The cost of resettlement needs to be looked at both in terms of the actual financial costs (surveying people and property, compensation or rebuilding of settlements and infrastructure and actual translocation), and also the human cost of the stress caused by uprooting. It is very hard to put a money value to this latter cost (indeed, you could argue that it is both foolish and inadequate to try to do so), but that does not make it unimportant. Hassan Dafalla describes the impact of resettlement on the Nubians displaced by Aswan: 'their ecology was shaken to the roots, and their environment presented new fields of experience and alien conditions to which they had to adapt themselves'.[5] It is difficult to imagine the pain of uprooting and extinction of place experienced by reservoir evacuees, but it is a very real downside of the 'development' process. The stress of resettlement is multidimensional, both psychological and socio-cultural. It is particularly grim that this stress is so often exacerbated by bad planning and inadequate provision for resettlement. Loss of assets, unfamiliar environments, unprepared resettlement sites, poor living conditions and hopeless economic prospects are all elements in the human and economic costs of resettlement.[6]

It is perhaps surprising that something as common as population resettlement should create so many problems, but it is true. Some African man-made lakes have been the subject of very detailed resettlement planning exercises, most notably Volta Lake in Ghana and the Kainji in Nigeria, both of them in the 1960s. However, even here, success was moderate. At Kainji, architect-designed concrete houses were judged uncomfortable and badly laid out, while at Volta Lake only 40 per cent of evacuees remained in resettlement villages in 1968. Yet these projects were done with care, and are usually regarded as fairly successful. Often the record is far less happy.

When construction of the Kariba Dam began, no resettlement studies had been carried out. The result was a 'poorly conceived and trauma ridden crash programme to get people out of the lake basin before the dam was sealed'.[7] While most evacuees were resettled nearby, in 1958, 6,000 Gwembe Tonga people were removed by force. In clashes nine people were killed and over thirty injured. When the Bakolori Dam in Nigeria was built, the property of the 12,000 people displaced by the reservoir was still being surveyed as the waters rose. Disputes about compensation rumbled on for several years, joining with other dissent on the project to generate the blockades to stop construction work on the irrigation scheme which led eventually to an attack by riot police in 1980 and a considerable death toll among farmers.[8]

Many reservoir resettlement projects are poorly planned and under-financed, and most evacuees are considerably worse off in the short term. In many cases their plight is permanent. The trauma of the resettlement process for Zambian Gwembe Tonga at Kariba, exacer-bated by the economic decline in Zambia since 1974, has significant implications for socio-economic conditions in the valley today. There can also be second-order problems for those living close to reservoirs, such as disease. River-blindness (onchocerciasis) is spread by biting flies whose larvae live in fast-flowing water such as that found in rapids and dam spillways. Bilharzia (schistosomiasis) spreads from an inter-mediate host in certain snails which like shallow water, and thrive on reservoir margins. Diseases such as bilharzia are widespread in African floodplains anyway, but dam construction can increase their preva-lence.

The reasons for the failure of much resettlement planning are rea-sonably easy to recognize.[9] The basic point is that resettlement is not some relatively simple procedure which can be tacked on to the back of a dam project without too many problems. It is extremely complex and sensitive, and it is remarkably difficult to do it well. It is quite an achievement to carry out a resettlement project which simply leaves no one disastrously worse off and excessively aggrieved.

Although there are many disciplines involved in designing and building dams, those necessary to get a good grip on resettlement are usually not very visible. Dam design is dominated by hydrology, engin-eering, geology and economics. Disciplines like sociology or anthro-pology or development studies are rarely given a professional role to play, and if they are it is too often a token one. They are brought in for too short a period when the project is already too far down the road towards completion. Too little time is allowed for the necessary surveys of existing property and possible resettlement sites, let alone any kind of participation. It is commonplace that too little money is allocated to resettlement to provide adequate replacements for what is lost.

One gain goes some way to compensate economically for loss of land below a reservoir and resettlement costs. This is the possibility of the development of a productive reservoir fishery. The ecology of man-made lakes in the tropics is complex, and received a great deal of research attention in the 1960s. The rate at which the reservoir fills and the fate of flooded vegetation are important, and there is substantial variation in the productivity of fish stocks and therefore the economic value of the fishery created.[10] At Kainji, many of the 5,000-plus fisher-men displaced by the reservoir were able to change their fishing methods and continue to fish on the new lake. More dramatic was the fishery which developed on the Volta Lake in Ghana, where in 1976 there were over 20,000 fishermen (many of them not evacuees, but fishermen from elsewhere), and the annual catch was some 40,000 tons. This gave a catch of just under 5 kg per ha per year. Indeed, the catch is more than four times that predicted before the dam was built. In this sense, the people of Ghana were very fortunate. Volta lake fish are a major source of protein in Ghana, and indeed given the fixed low price at which the electricity generated by the dam is sold, this fishing industry rates as one of the most significant economic benefits of the dam.

However, it should be noted that the success of the fishing industry of the Volta Lake cannot necessarily be expected from other reservoirs with different ecologies. Studies of seven man-made lakes in Africa done in the 1970s showed that catches varied from over 12 to under 0.8 kg per ha per year.[11]

## Dams and floods

Floodplains are environmentally complex, and so are patterns of human use. The subtlety and sophistication of indigenous adaptation to patterns of flooding, floodplain soil conditions and resources outside the floodplain have been described in Chapter 4. Many different aspects of floodplain economy are affected by dam construction.

The most basic impact, of course, is on the pattern of river discharge itself. Dams store water, and as such inevitably affect the magnitude and timing of downstream flows. Floods are delayed and reduced in magnitude by the construction of a dam; in other words, flood peaks are smaller although flood flows may be longer-lasting The exact effect depends on the size of the dam and reservoir and the way they are operated. These in turn depend on the purpose of the project. Dams to store irrigation water for dry-season use, like that at Bakolori, are often filled early in a wet season before water is released downstream. Such dams can change the regime of a river from one with a short flood season with very high-magnitude flows, into a river with a shorter flood season with more moderate flows and some residual flows through the dry season. Even once full, such dams continue to affect flood flows downstream, because floodwater entering the reservoir is delayed in passing through it. This 'flood-routing' effect lowers and delays flood peaks. Dams for flood-control maximize this dampening effect on river flows. Dams for hydro-electric power (HEP) tend to create a fairly continuous annual pattern of flooding. However, peaks in power demand can affect flood flows. Analysis of discharge in the Tana River in Kenya, on which a series of HEP dams have been built, shows a weekly cycle, with higher flows at weekends related to reduced electricity demand in Nairobi.

In many cases, dams nominally have a multiple purpose role, combining two or more functions. In practice, one function tends to dominate operation, if only because the different uses to some extent conflict. Although in theory specific plans could be made to release water for downstream users, in Africa this has very rarely been done. In practice dams are operated in isolation from the needs of downstream users. The consequential 'side-effects' can be very significant.

Most of the major African rivers have been dammed in at least one place, and flow in many of the largest rivers is controlled over much of their length.[12] Both tributaries of the Nile are dammed, and the dam at Aswan controls flow of the combined river within Egypt. In the Zambezi basin there are dams on the Kafue at Kafue Gorge and Itezhitezhi and on the Zambezi itself at Kariba and Cabora Bassa. In West Africa, the Volta River is dammed at Akosombo, the Bandama (Ivory Coast) at Kossou, the Senegal at Manantali and Diama. There are numerous dams in the Niger basin, including the Bakolori on the Sokoto and the Lagdo on the Benue. There are numerous dams on the Niger itself, upstream at Sélingué and Sotuba for HEP and Markala and Karamsasso for irrigation, with more planned in the headwaters and on the Bani River as well as further downstream in Nigeria (Kainji and Jebba). It is likely that storage in the near future will amount to 12 per cent of dry-year inflow to the Niger Inland Delta. Within the Chad basin the Hadejia-Jama'are river system is also extensively dammed.

Other river engineering projects also have significant effects on wetlands. In the Sudan the Sudd may be extensively affected by the Jonglei Canal, and in Botswana the Okavango Delta by the dredging of the lower Boro River as part of the Southern Okavango Integrated Water Development Scheme,[13] as well as by upstream diversions in Namibia. There is a similar proposal to carry water past the Hadejia-Nguru Wetlands in northeast Nigeria by channelizing the main rivers. Dams are just one of a range of major engineering projects that affect African rivers.

The effects of dams on river flows of course varies. Given the limited amount and poor quality of hydrological data available through much of Africa, and the prevalence of drought, it is no mean feat to establish the exact nature and extent of hydrological impacts. However there is no doubt that the impact on seasonal rivers can be very marked. Prior to construction of the Akosombo Dam, discharge in the Volta fell to a trickle in the dry season, and the seasonal peak was high, and very variable. In 1957 it reached 320,000 cubic metres per second. The following year it was only a quarter of this. Annual discharges fell very low between 1964 and 1968 as the dam filled. Once complete, the seasonal pattern was much less extreme as water was stored and passed through the hydro-electric turbines through the year, although drought in the late 1970s and early 1980s caused the level of the reservoir to fall to such a level that HEP generation was curtailed.

The River Benue, which flows from Cameroon into Nigeria to join the River Niger below Kainji, was dammed in 1982 above the city of Garoua. The Lagdo Dam impounds a fairly shallow lake of 70,000 ha, and is intended to generate 300 GWh of HEP for Garoua. Garoua lies below the junction of the Benue and another river called the Mayo Kebbi, and hence enjoys both regulated and unregulated flows. Nonetheless, when the Lagdo Dam was closed to fill the reservoir in 1982, discharge of the Benue at Garoua fell to only 43 per cent of average flows, even though it was a wet year.[14] The flood also lasted only 3–15 days, compared to a normal flood of 93–160 days, and flood depth was reduced to a quarter of pre-dam levels. 1983 was a year of low rainfall, so although the reservoir was virtually full, floods were still below normal. In the mid-1980s there was still an allowance for downstream releases of 2.7 billion cubic metres. This is only 35 per cent of pre-dam flows, but even this will be eroded through the 1990s as demand for electricity rises and year-round water releases are increased at the cost of peak flood discharges.

What effects do these changes in natural river hydrology have? First, they affect the ecology of natural floodplain environments, and hence human use of resources such as grazing. Secondly, they directly affect the use people make of floodplain environments through agriculture and the economic value of their production. Thirdly, they affect the breeding of fish, and hence fishing economies. Fourthly, they affect the balance of resource uses, and create the context for disputes between different groups of people for a shrinking resource. Each of these will be discussed in turn.

## Dams and floodplain ecology

Vegetation communities and wildlife populations in the floodplains of seasonal rivers in Africa are adapted to and dependent on natural patterns of river flow just as people are. River control and transformation of those flooding patterns have serious implications for floodplain ecology. This is of more than esoteric interest, since very often people depend either on these resources themselves or on the same flow patterns. As so often, wildlife provides a model of what environmental change is doing to people. It is bizarre that through wildlife television and international organizations, sympathies and attention in the First World are often more swiftly and strongly engaged by the plight of wildlife than of people, but such concern may be better than none at all.

This interplay between impacts of river flow changes on wildlife and on people is well demonstrated in the case of the development for irrigation of the Logone floodplain in northern Cameroon.[15] Two polders (embanked areas within the floodplain) for irrigated rice cultivation have been built by the Société d'Expansion et de Modernisation de la Riziculture de Yagoua (SEMRY). These total about 12,000 ha, and could produce 7,000 tonnes of rice (70 per cent of Cameroon's consumption). The first of these, SEMRY I, covers 5,200 ha and takes its water direct from the River Logone. The second, SEMRY II, (6,600 ha) takes water from a 39,000 ha reservoir built in the floodplain (although there is also provision to take water from the Logone). This reservoir, Lac Maga, was built in 1979. It takes water flowing into the floodplain from local runoff, and with the dykes built within the floodplain this has a significant effect on flooding. It is estimated that in addition to the area flooded by the reservoir, flooding is prevented on about 59,000 ha, and affected over a further 50,000 ha. Lac Maga fills from August onwards with water falling on the Mandara hills, capturing more than half this water (200 million cubic metres) in normal years. Flood embankment, plus the offtake of water to fill Lac Maga and to supply SEMRY I, delays and reduces flood peaks in the Logone itself, and hence affects the timing of inundation downstream.

One of the richest National Parks in West Africa, Parc National de Waza, lies downstream of the SEMRY Projects. The productivity of its floodplain grasslands depends on seasonal flooding. Following completion of Lac Maga in 1979, flooding was very poor in Waza. In 1980 elephant and other wildlife were found in a poor state around dried pools in the floodplain, and in 1982 a campaign was launched to dig pools, build canals to bring water to the floodplain and even to truck in water by lorry. Studies by the Delft Hydraulics Laboratory suggested that it was drought and not the reservoir that was at fault, and plans were made to supply water to Waza by building a canal from Lac Maga, or by improving flooding from the Logone. Such investment in the Parc National de Waza could, according to Drijver and Marchand, be justified 'by its profound natural and touristic value'; however, they also note that 'it is remarkable that, in finding solutions to the drought of Waza, and talking about canals and artificial freshets, no attention is paid to the possibility of improving hydrological conditions for traditional herding, recession culture and fisheries downstream of Lac Maga'.[16]

The picture here is complicated, because the area exposed as Lac Maga rises and falls each year is likely to provide good grazing which will go some way to offset the losses elsewhere, and the shallow lake may also support a productive fishery. Nonetheless, there is a net loss of grazing and some adverse impact on downstream sorghum cultivation. The area under seasonally flooded perennial grasses *Echinochloa pyamidalis* and *Vetiveria nigritana* has fallen, as has the productivity of what remains.[17] No rice has been planted in Kotoko villages since 1982. The alteration of flooding patterns will require considerable adjustment on the part of floodplain communities. The impact of drought greatly increases the potential impact of human-made control structures, and (perhaps more seriously) greatly increases the level of uncertainty in river management.

In the lower reaches of the Tana River in Kenya a narrow strip of evergreen floodplain forest is maintained by flood-recharged groundwater in an area otherwise too dry for more than Sahelian savanna vegetation.[18] In its turn, this forest supports populations of two rare and localized subspecies of monkey, the Tana River Red Colobus and the Tana Mangabey. The forests are also used extensively by Pokomo and Malekote people who farm along the river banks. The Tana River flows for about 650 km from the slopes of Mount Kenya to the Indian Ocean, and was dammed in several places in the headwaters between the 1960s and 1980s. There are two flood periods each year on the Tana, and great variability between years. It seems that the occasional very large floods are important for the establishment of new forest trees. Dam construction, particularly the closure of the Masinga Dam in 1982, has reduced the size of the largest floods, and it is predicted that in the long run the forest will cease to replace itself. Given the other pressures on the forest, such as the cutting of trees for charcoal production and building timber for settlers in new irrigation schemes along the river, the future of the forest looks grim.

Similar ecological impacts are predicted as a result of the construction of the Turkwel Gorge Dam in Northern Kenya.[19] There is a very thin strip of forest dominated by *Acacia tortilis* along the River Turkwel, maintained by a shallow water table supplied by periodic river flow. The depth of the water table is critical to the survival of the trees, and again the occasional very large floods are essential to their regeneration. The dam in the Turkwel Gorge is intended to supply HEP to Western Kenya. It will transform the Turkwel's highly seasonal flow regime into a more continuous flow, and in doing so will reduce the size of flood peaks. The sandy river bed and massive evaporative demand is likely to mean that little if any water reaches the forests of the lower river. These forests are of more than ecological interest, since they provide essential dry-period browse resources for Turkana livestock. The nutritious seed pods of the Acacia are particularly important. Without the forest, the entire structure of Turkana resource use is threatened. On a more mundane level, the water supply to the District headquarters at Lodwar also depends on the shallow floodplain acquifer, and is also threatened. The contract to build the Turkwel Gorge Dam was taken in 1986, after a rather surprising shift in aid donors from the European Community (who were insisting that an Environmental Impact Assessment (EIA) was carried out) to a consortium of French commercial banks. No EIA was carried out. The failure of the donor community in this instance to make sure an EIA was carried out is an important indicator of the real pressures on the project approval system. This problem is discussed further below.

A number of African wetlands on seasonal rivers are characterized by extensive seasonal grasslands which comprise important grazing resources for pastoral groups. Seasonal movements by pastoralists into floodplains in the dry season can involve very large numbers of animals. There are probably 1–1.5 m cattle plus one million sheep and goats and 0.7 m cattle in the Niger Inland Delta in Mali.[20] The *toich* grasslands of the Jonglei area in Sudan hold between 0.5 m cattle in the wet season and 0.8 in the late dry season (plus between 0.1 and 0.2 m sheep and goats.[21] In the dry season the Kafue Flats in Zambia support about 0.25 m head of cattle, half the population on the Kafue Basin as a whole.[22] Other surveys record 0.15 million cattle on the Shire marshes in Malawi, and 0.3 million in the floodplain of the Gambia River.[23] Where river control makes significant changes to seasonal flood regimes there can be serious repercussions on pastoral groups.

Dam construction on the Kafue River in Zambia has proceeded in several stages. In Stage 1, completed in 1972, the Kafue Gorge Dam was built. This held four 150-MW HEP generators, and impounded a reservoir of 800 m$^3$. Stage II, completed in 1978, involved construction of the Itezhitezhi Dam to provide back-up storage and two more generators to raise total capacity to 5,000GWh. The Itezhitezhi dam formed a much larger reservoir of 5,700 m$^3$.[24] Moreover it lay upstream of the Kafue Flats, an extensive area of seasonally flooded land. The combined reservoirs were intended to provide a guaranteed flow of 183 m$^3$/sec at Kafue Gorge, of which 15 m$^3$/sec was allocated for non-HEP purposes. Each year it was planned to fill the Itezhitezhi reservoir, and then release surplus flows downstream. When the operating rules established for the upper project were applied to flow data for the years 1951–75 it was estimated that there would have been be no spill from Itezhitezhi one year in five (five years out of twenty-five). The only releases in dry years were therefore expected to be those for HEP generation, plus the 15 m$^3$/sec allowed for other purposes. In this circumstance, it was planned to release a freshet of 300 m$^3$/sec 'as required by ecological considerations'.[25] In the event, a study of flood extent using 1981 satellite data showed fairly complex impacts on the extent of flooding, with some areas wetter than before but a reduction in the duration of floods and erratic fluctuations in water levels in the dry season.[26] Already by 1978, local people were complaining that cattle had been drowned because of unexpected releases of water from Itezhitezhi.[27]

Potentially the most serious single river engineering project in terms of direct impacts on pastoralists is the Jonglei Canal in the southern Sudan (see Figure 6.3). The Sudd region consists of a vast area of permanent swamp (covering some 18,000 square kilometres) and seasonal wetlands (an additional 11,000 square kilometres) flooded by the Bar El Jebel. In addition to the swamps proper, there are extensive grasslands flooded by the river (*toich*), grasslands flooded from local runoff and woodlands.[28] Almost half of the flow in the White Nile at the South end of the Sudd (about 50 billion cubic metres) is lost in its passage through the area, and as described in Chapter 5, planners have for many decades been attracted by the notion of building a canal to bypass the Sudd and thus getting more water for irrigation in Egypt and northern Sudan. That idea was one of the elements of the core of the Equatorial Nile Project, and although the idea of a Jonglei Canal was not pursued at that time, it was brought up again in 1974. Work began in the late 1970s.[29]

The Sudd region contains 2–400,000 people. Most of them are Dinka and Nuer, who are essentially pastoralists who move from villages on higher land down into the grasslands as the floods recede. They also grow crops in the wet season (May–November) around their villages. The Shilluk live on higher land and are more sedentary, with greater emphasis on agriculture. There are some 0.8 million cattle and perhaps 0.2 million goats and sheep in the area, plus about half a million wild grazing animals, including 0.36 million tiang, a large antelope.[30] Flooding of grasslands begins with local rains in April, then the rising level of the Nile brings river flooding into play through until the end of the year. Livestock are moved from rainfed grasslands to lower areas as the floods recede. The wild tiang also migrate in response to flood patterns.

The Jonglei Canal was designed to be 360 km long and to carry 20 million cubic metres of water per day. Construction began in 1978,

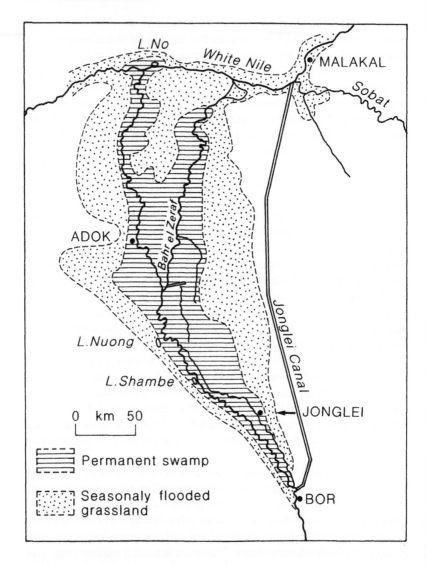

**Figure 6.3** Jonglei Canal and the Sudd, Sudan

but work has been halted since the early 1980s by civil war. It is not therefore possible to do more than predict its ecological and socio-economic impacts. However, an extensive study was carried out in the late 1970s and early 1980s.[31] Predictions are made more complex by an apparent shift in the flooding regime of the White Nile and resulting uncertainty about the proper basis for analysis and prediction. The level of Lake Victoria, and its outflow in the White Nile, rose after

1961–4. In retrospect it is clear that discharges were low from 1901 to 1961 and much higher from 1964 to the early 1980s. A survey using satellite imagery of the area flooded in the Sudd in 1973 suggested that it was double that in the 1930s. Subsequently, of course, Nile flows fell to a low level in the mid-1980s. There is also uncertainty about likely canal operating procedures. However, attempts to simulate the effects of the canal on floods suggest a decline in the area of permanent swamp of between 32 and 43 per cent, and in seasonal swamp of between 11 and 32 per cent.[32]

Clearly, if it were finished, the Jonglei Canal would bring about a substantial reduction in flooding in the Sudd, and ecological changes over large areas. There would be reduced flooding west of the canal and a substantial decline in the area of both swamp vegetation and the area of seasonal flooding. There would also be changes in the extent of different kinds of grassland which would favour less nutritious species and considerable impacts on fish populations that might involve considerable declines in the productivity of the fishery. At the same time, there would be likely to be some new flooding to the east of the canal where local runoff from the southeast is prevented from reaching the Sudd by the canal banks, and this may in time support grassland, and perhaps a fishery.

There would also be problems with access across the canal. It would be more than 50 metres wide and fast-flowing. Offset ramps have been designed so that cattle can swim across and (theoretically) find a ramp waiting for them when they get to the other side. It is estimated that about a quarter of a million people and some 0.7 million cattle will have to cross every year, as will very large numbers of wild herbivores.

Three quarters of the Jonglei Canal had been dug by 1984, when work was halted as a result of the renewal of civil war in southern Sudan. Thayer Scudder comments that 'it is hard not to believe that Jonglei Canal was not a factor in the 1983 renewal of this present strife'.[33] Certainly it has become a theme within it. Construction equipment has now deteriorated, perhaps beyond repair, and there are reports of siltation and erosion damage to the canal structures that have been built. It would be remarkable if the project were completed in the near future, even if hostilities end permanently. The cost of completion would be vast. In addition, the value of the canal in terms of Egypt's water needs is now being re-evaluated. Its large flow amounts to only 4–5 billion $m^3$ per year, compared to the 93 billion $m^3$ annual discharge at Aswan. Jonglei may postpone Egypt's water shortage, but it cannot solve it.

# Dams and floodplain farming

One of the most important impacts of river control on floodplain economies in Africa is the disruption of floodplain agriculture. Two separate factors are at work: firstly changes in water quality and secondly
changes in the amount of water, specifically the amount and timing of
water flooding patterns. Of the two, there is far less information about
the effect of dams on water quality. However, from studies of the problem of reservoir siltation caused by soil erosion it is clear that dams trap
sediment being carried by rivers. Silt has accumulated very rapidly in
the Khashm el Girba Reservoir on the Atbara River in Sudan that
supplies irrigation water to Nubian evacuees displaced from the Aswan
Dam. By 1977 the original storage capacity had dropped by 59 per
cent.[34] The trapping of silt within a reservoir can rob downstream areas
of fertility. It is argued, for example, that the trapping of silt by the
Aswan High Dam has caused reduced fertility in areas once flooded
annually in the Lower Nile Valley, as well as erosion of the Delta because less sediment is being brought down to it. Aswan was primarily
built to allow double-cropping and an extension of the irrigated area
in Egypt. It was completed in 1969, and designed to have enough spare
capacity to store the expected silt load for the next 500 years. Apart
from reductions in soil fertility, water released from a reservoir which
has lost all or most of its sediment can be highly erosive, and cause the
loss of riverside land, as has happened below the Kariba Dam. There
can in some instances be problems of poor water quality in reservoir
releases, for example where releases are dominated by water rich in
hydrogen sulphide released from deep within a reservoir.

Evidence of impacts of reduced water supply to floodplain agriculture is better understood. In the Benue floodplain, for example, the
reduction in flooding caused by the Lagdo Dam has had a significant
impact on flood-recession farming with sorghum. In the dry years of
the mid-1980s, the reduction in area was 50 per cent,[35] and there was
disruption of production over a wider area because of the unpredictability in timing and height of flooding. The Bakolori Dam on the
Sokoto River in Nigeria had similar impacts on floodplain agriculture
between the dam site and Sokoto City and the confluence with the
River Rima (about 120 km).[36] Closure of the dam affected the timing
of flooding downstream, and reduced both its extent and depth. This
meant that farmers no longer knew what to expect from the flood, and
their ability to match expected flooding, soil and crop type broke
down. Rice cultivation in the wet season became more risky than before, and there were areas too dry for rice and yet too easily waterlogged by rainstorms to support dryland crops such as millet. More

land was therefore left uncultivated in the wet season (five fields out of ten compared to two out of ten before the dam), and there was a significant reduction in the number of fields growing rice.

There was also a major decline in dry-season irrigation in the Sokoto Valley. Reduced flooding meant more limited opportunities for farming beyond the wet season using residual soil moisture as well as reduced recharge of the shallow aquifer under the floodplain. This meant that wells had to be deeper to reach water. This meant that they were more difficult and more costly to dig, and that more labour was required to water gardens. In some cases the aquifer was too deep, and wells could not be dug in loose floodplain sediments. In villages surveyed in the early 1980s, up to three quarters of dry-season irrigators had given up. In general only the larger operators survived. The dam reduced the area of rice by 7,000 hectares and dry season crops by 5,000 hectares, out of a total of 19,000 ha of floodplain land.

Luckily for the Sokoto Valley farmers, small imported petrol pumps began to become available in the mid-1980s, and a State 'Integrated Agricultural Development Project' with World Bank funding began a shallow tubewell digging programme.[37] This appears to have gone some way to compensate for the loss of the natural flooding of the Sokoto, but obviously at some cost. In terms of economics, the downstream impacts of the Bakolori Dam were serious, because the losses of production in the floodplain make a significant difference to the economics of the irrigation scheme itself. Of course, the designers of the Bakolori Project did not take the downstream impacts into account – but they should have done. Had the proper surveys been done it is likely that the dam and irrigation scheme would never have been built. This would have been very good news for the floodplain farmers, and also (as it turns out) for the farmers of the irrigation area and the Federal Government's budget. But that is a story that must wait until Chapter 7.

## Dams and floodplain fishing

The close adaptation of the life cycle of fish of floodplain rivers to annual flood patterns has been described in Chapter 4. Droughts in the 1970s and 1980s had marked effects on the fish populations and fish catches, for example at Mopti in the Niger Inland Delta.

Significant impacts on fish populations are therefore to be expected as a result of changes in annual flow regimes brought about by dams, and indeed a number of studies have shown this to be the case. Work on the Niger River downstream of the Kainji Dam in the 1960s showed that catch sizes fell and there were changes in the composition of the fish population.[38] The number of seasonal fishermen also fell in the Niger Valley, although there was good fishing immediately below the dam in the tail-race water. Similarly, in the Sokoto Valley in Nigeria, fishermen complained of reduced catches following closure of the Bakolori Dam in the late 1970s, and a number had taken to travelling hundreds of kilometres to fish elsewhere, for example at Lake Chad.

In the floodplain of the Pongolo River in Natal, South Africa, a number of fish failed to spawn because dam construction reduced flood peaks to such an extent that there was no access to floodplain pools. In Ghana, the clam beds and hence the clam fishery of the brackish waters of the lower Volta moved in response to the variations in river flow during the construction of the Akosombo dam.[39] However, in this instance the fishery survived, at least until the droughts of the 1980s. In the Rufiji Delta in Tanzania, the fin fish and prawn fishery is estimated to have a value of US $32m per year. A number of dams have been proposed on the Rufiji, and although none have been built to date the possibility remains open. Such dams would have a significant impact on sedimentation and salinity in the delta, and hence on the ecosystems that support the fishery. Reclamation of mangrove areas for agriculture would also have a serious economic impact.[40]

It is predicted that the Diama Barrage on the River Senegal, built to prevent incursion of salt water during periods of low river flow, will cause the loss of some 7,000 metric tons of shrimps and fish in the Senegal Delta. There are likely to be much larger losses (about 360,000 tonnes) in the Middle Floodplain of the Senegal due to the flood-control effects of the Manantali Dam and the development of dyked irrigation schemes. These losses are almost three times the predicted gains from new fish resources in reservoirs within the basin.[41] Of course, this balance need not be negative. A similar comparison, this time of predicted gains and losses in fish production, has been drawn up for the Benue River upstream and downstream of the Lagdo Reservoir in Cameroon, and here the balance is slightly positive (a net gain of 2,000 tons a year).

Dams also affect fish populations of temperate rivers. Most significantly, they can block the migration routes of valuable game fish such as salmon. The common response is to install some kind of fish-pass, but this is rarely done in Africa. Those built have not proved very successful.[42] The Markala Barrage in the Niger Inland Delta, built to supply the Office du Niger, acted as a major blockage to upstream movement of fish, and there was a significant decline in the fishery upstream. A fish ladder was installed, but was ineffective, partly because it was too small, and partly because while fish move extensively in floodplain rivers the long upriver spawning migrations of temperate game fish do not occur. It is also likely that floodplain river fish would need very differently designed fish ladders from those built for temperate fish. The research necessary to know what is needed has not been done.

## Dams and floodplain resource conflicts

The reduction in the extent of flooding caused by dam construction tends to reduce the extent and productivity of floodplain resources. One effect of this can be the emergence of conflicts over the resources that remain. Dam construction is only one of a range of possible triggers to such conflict. In 1979, conflict over floodplain land in the Senegal Valley reached particularly serious proportions.[43] It was the spark for extensive inter-ethnic violence between black Senegalese farmers and light-skinned 'Moors' who form the political élite in Mauritania in 1989. At least 200 Wolof and Peul were killed in Nouakchott, and 35 Moors in Dakar. The Senegalese government placed 15–20,000 Mauritanians under military protection, and there was eventually an international airlift in both directions of some 180,000 people.

While this is a long-standing conflict, it was exacerbated by the takeover of land by the Moorish élite for irrigation in the 1980s. The dam at Manantali made modern irrigation suddenly an attractive target for investment. This led to the loss of floodplain pastures and agricultural land for the black communities in the valley. At the same time drought and changes in river flows because of the dams upstream led to increasing pressure on the remaining floodplain. This led in the 1980s to conflict between farmers and herders from both sides of the river. It was such a conflict between Mauritanian herders and Senegalese farmers in the river bed that sparked the first violence in 1989 and which eventually led to the ethnic fighting in Mauritania and Senegal. This violence reflects therefore both the new opportunities for irrigation, and the problems of drought faced by pastoralism on the Saharan fringe.

These changes have been supplemented by pressure by both multilateral and bilateral donors to 'roll back the state' in Africa, and let private enterprise and an unfettered market play a larger role in economic change. Partly to this end, a new Land Law was passed in Mauritania in 1983 that legalized private ownership of land. Under this law, the government would grant title to people with the resources to develop land. This strongly favoured élite groups and would-be capitalist farmers, predominantly from the Moorish minority. In Senegal a new agricultural policy in 1984 promoted commercial agricultural development. In both countries the result has been rising demand for flood-recession land by groups outside the floodplain, and an effective 'land grab' at the expense of local people. The river does not divide different ethnic groups, indeed members of the same families sometimes farm both banks of the river. The land grab has been particularly abrupt in Mauritania, where the government evicted 'foreign' (i.e. black) farmers. Senegal responded by closing its frontier to livestock, and so the conflict began to escalate. 'Privatization' is a cover for the accumulation of landholding by the Mauritanian élite. As Michael Horowitz comments, 'the land tenure system being replaced is hardly egalitarian, but it is characterized by broad access to productive lands on the part of a large number of smallholders and tenants, and the lands being expropriated in the name of privatization are already privately owned'.[44]

Similar, although much more localized, conflicts about shrinking floodplain resources are emerging in the floodplain of the Hadejia and Jama'are rivers in north-eastern Nigeria. Where these rivers join there is a large area of floodplain wetland mixed with areas of dry land. Some of this is permanently flooded, other parts flood when the river rises every wet season. This area has long been used by Fulani pastoralists, particularly in the dry season when there is a substantial rise in livestock numbers.

The floodplain is vital to the annual cycle of movements of the Fulani because rainfall of the area around it is low (about 500 mm) and highly variable. The rains were particularly poor in a number of years in the 1970s (1972–1978) and 1980s (1980–1987).[45] The wet season is short, and the river regime is flashy, with 80 per cent of runoff in the Hadejia and Jama'are coming in August and September. The timing, extent and duration of flooding in the floodplain varies from year to year, but has suffered a significant decline in recent years because of drought and the construction of the Tiga Dam in the 1970s to supply irrigated land upstream. In 1950 the flooded area is estimated to have been 2,350 square km. In 1986 it was 1,186 square km and in 1987 only 700 square km. This shrinkage of the wetland has occurred over the same period as the intensification of demands for its use. Upstream dam construction has increased the pressure on the floodplain, and the tension between its people.

The Hadejia-Jama'are floodplain is also now subject to increasing agricultural intensification, particularly through the widespread adoption of small pumps for irrigation. This is allowing land to be cultivated in the dry season which would once have been open for cattle grazing, and there is a serious conflict emerging between pastoralists and farmers. There were a number of pitched battles in the late 1980s, and a number of people have been killed. Conflict over land is revealing fundamental changes in the communal basis of land tenure, the mechanisms for allocating land and the legal procedures for deciding conflicts. Among other pressures is the development of quite large irrigated wheat farms by 'élite' farmers (including urban businessmen and army officers) in the wetlands.[46] A ban on wheat imports has made such farming highly profitable. Under the Nigerian Land Use Act an urban capitalist farmer can obtain a statutory right of occupancy to rural land presently used informally by local people. The result is an effective land-grab of irrigable areas, and further pressure on increasingly scarce resources.

The scale of debate about the future of the area is now expanding rapidly, because as the rivers carry smaller floods, areas further downstream are also experiencing lack of water, particularly for irrigation. These areas lie within Borno State, much closer to the state capital of Maiduguri. Further downstream, the combined river (the Komadugu Yobe) forms the international boundary with the Republic of Niger, and farmers in Niger are also affected by reduced river floods. The lack of water has provoked minor border incidents, high-level meetings between the Borno and Niger governments, and meetings with Kano State to devise ways to increase the flow of water downstream.

One of the more alarming proposals, being investigated by the FAO in 1989, envisages digging a wholly new river channel through the wetlands of the Hadejia-Jama'are junction to carry water past the area with minimal evaporation losses. This would be rather like the Jonglei Canal, and would have similar effects on grazing resources, and indeed on the feasibility of wetland irrigation. A less extreme plan is to simply dredge out channels within the wetland and to build sluices to direct more water to downstream areas. Such engineering has already been tried by the farmers themselves. Farmers who lack water have come in lorries to the wetter parts of the wetlands to try to dig out old channels, and there have also been cases of bunds being forcibly thrown down. Those in areas which have been becoming drier blame small-scale rice polders built in wetter areas to improve flood control and the quality of rice husbandry. It is a classic response to a problem caused (in part) by engineering projects to seek a solution through the construction of more engineering works. It is not necessarily a very sensible response. The environmental implications of river diversion structures and channelization would need very careful study, and would be by no means easy to predict. The political implications of interfering with the distribution of water in the Hadejia-Jama'are Valley are considerable.

The problem of a confused legal position over land and the constant possibility of conflict in northern Nigerian floodplains bear a close resemblance to the situation in the Niger Inland Delta in Mali described in Chapter 4. Like the tragic violence in the Senegal Valley, these are by no means isolated instances of conflict over floodplain resources in Africa. Similar conflicts are arising in many places as pressures on wetland resources increase.[47] In many cases the immediate pressure has often been drought, sometimes combined with investment in major infrastructure such as dams. Behind these triggers of conflict lie deeper-seated problems of the erosion of indigenous management systems.

Conflict over resources, and the degradation that often results, is not caused by a failure of indigenous management systems so much as by their dislocation by political and economic forces associated with state formation and development. The range of possible responses to such pressures currently on the table are few. In the case of the Niger Inland Delta, Drijver and Marchand comment, 'future conflicts . . . can only be prevented by means of an integral river basin-wide development plan'.[48] The limitations of such planning have been discussed in Chapter 5, but it may be the best prospect for the future. Certainly, without the discipline of integrated planning, large-scale externally-conceived water development projects can have a serious and destabilizing effect on the complex patterns of floodplain resource use.

## Dams and planning disasters

It should be clear by now that too many of the dams built on African rivers have had serious, and in some cases near disastrous environmental impacts. Furthermore, these impacts have often not been predicted, and even once they have occurred they are very often not fully understood. Certainly they are rarely dealt with in any adequate way. I have gone into considerable detail about the nature and extent of these impacts not because I enjoy parading planning failure, but because I believe that it is important that the impacts of dam construction on the complex and inter-connected nature of floodplain resource use is properly understood. The assumption that floodplain resources are under-used, and that natural environmental processes and patterns of human use are well enough understood that dams can be built without adverse environmental impacts, are the root cause of many of the numerous failures of African river basin development.

The ideology of river basin development discussed in Chapter 5 offers a number of possible reasons why planning failures persist. It is perhaps less easy to see why Northern technical planning seems so incapable of realizing its self-image and predicting environmental impacts and taking them into account. There is enough money spent by aid donors and African governments on consultancy reports by supposedly excellent First World firms. In theory, there are planning methods which should be capable of taking environmental impacts on board and leading to decisions about development that avoid disasters. Why don't they work?

The basic tools used in project decision-making are Environmental Impact Assessment (EIA) and Cost-Benefit Analysis (CBA). Both are well-established, and have been the subject of repeated treatises, conferences and 'hands-on' manuals.[49] EIA techniques involve attempts to identify the nature and severity of environmental impacts of development projects, and the significance of those impacts on people. They may lead directly to recommendations that a project be developed or abandoned, recommendations for ameliorating measures and measures for monitoring future environmental change. Even in First World countries, EIAs are often controversial, failing to satisfy environmental critics. There are various reasons for this. Often they are begun too late, so that their results cannot affect the decision to go ahead with a project. They are also very often carried out in a watertight box, such that there is no formal mechanism for integrating results into decision-making. Even where EIAs are demanded by law, notably under the National Environmental Policy Act passed in the USA in 1969, they are not always useful tools for decision-making. It is argued that one tactic of those promoting project development is simply to make EIAs so voluminous that no clear and concise solutions emerge. Where EIAs are not mandatory, the decision to carry one out can itself be controversial and highly-charged. Many developers feel that the whole procedure is a waste of time and money.

All these problems exist in Africa, plus a few more. One of the most critical is the complexity of floodplain resource use systems and the importance of secondary and even tertiary environmental impacts. Impacts can be felt in locations remote from a dam site, and can be delayed by months or years. Impacts can also be cumulative, with the effects of one project being compounded by those of another, or by natural changes such as variations in rainfall. It is is extremely difficult for anyone to make accurate predictions about such impacts, particularly given the lack of information about the environment which is usual in Africa. Hydrological records are usually short, ecological surveys have rarely been done, and there are rarely scientists on hand with experience in the area affected. As a result, once again, it is overseas consultants who end up being paid to carry out EIAs, with all the problems attendant on the budget-conscious work of expatriate 'experts'. The teams carrying out such work often lack ecologists, sociologists and other disciplines, particularly if the EIA is not seen as an important separate item of the appraisal process but is integrated more loosely into discussions of future costs and benefits. Economists, hydrologists and engineers from Europe or North America are not the ideal people to carry out EIAs on African rivers.

EIAs are often done too late to affect the design of development projects. T. O'Riordan points out that 'it is a fact of contemporary life that environmental safeguards and local community wellbeing come late in the stage of project design and site selection'.[50] It is argued that there is now a much greater awareness of the need to include environmental appraisal routinely in feasibility assessments, both on the part of aid donors and the commercial companies they employ.[51] In 1989, for example, the World Bank adopted a new policy on dam and reservoir schemes which was 'designed to ensure that all environmental aspects of such projects become routinely and systematically integrated into project design and operation'.[52] There is a big step between policy formulation and the reform of practice, but such policies are a help. It is vital that there is a much more detailed analysis of what natural systems can contribute (and are already contributing) to socio-economic development at the earliest possible stage in project formulation.

EIA is often seen as the most important way of taking account of environmental aspects of dam construction, but it is less central to decision-making than the economic assessment of Cost/Benefit Analysis. CBA basically attempts to compare future income and expenditure from a project in order to see how they balance out over time. In theory, if all the benefits expected from a project do not outweigh all the costs by a significant margin, the project should not go ahead. Economists usually look for what is called an 'internal rate of return' (IRR) of about 10 per cent. Usually, they manage to find it.

There are all sorts of creative accounting which can be done to suggest that an acceptable IRR or cost-benefit ratio is possible. The simplest is to under-value costs and over-value benefits: simply to be unduly optimistic. This is not unusual, and its impact on CBA calculations for irrigation schemes will be discussed in Chapter 7. In the case of dams at least, failure to predict costs (and benefits) accurately may simply reflect ignorance of floodplain environments and people. If those planning a project do not have the technical knowledge to understand floodplain environments (for example if their team lacks ecologists) or floodplain resource use (e.g. if the team has too small an input from sociologists or anthropologists) it is unlikely to be able to assess the impacts of a dam properly. If in addition the terms of reference of their contract do not stipulate that they should consider downstream or other remote impacts, they will have no reason to look beyond the end of their nose. Indeed, any attempt to do so is likely to be highly unpopular with the client or with senior management of the consultancy company (who may not want to hear about difficulties that might delay the project or make it more expensive). Lateral thinking of this kind is likely to be expensive, and to be impossible within budget. The combination of ignorance (compounded by disciplinary biases within the planning and design team) and the straightjacket of the contract very often explains why appraisals are too narrow and are made too late to provide a sound basis for decision-making.

There are other problems with the way those setting the objectives of CBAs tend to treat environmental costs. Costs and benefits come on stream at different times in a project's life. The costs of construction, for example, are concentrated very early in a project's life. Economists consider costs (and benefits) that occur now to be greater than those which occur in the future, that is they discount streams of future costs and benefits to the present. The logic of this is that money invested now earns interest, and is therefore worth more in the future. So future costs and benefits are (in comparison) worth less than those of the present. The proportion by which they are less is called the discount rate. The adoption of different discount rates can have a significant effect on the apparent economic viability of a project. Whatever happens, values more than ten years down the line are usually more or less irrelevant in the cost/benefit equation. They certainly exist, and they often matter a very great deal in human terms, but they are not taken into account.

A further problem is that costs and returns are often aggregated in CBAs. This means that the method does not take any particular account of *who* bears the costs. Thus a dam project might generate significant economic benefits in urban areas through the generation of electricity, and significant costs locally in the loss of floodplain farming and fishing. If these two are simply placed against each other in a balance sheet, the benefits might outweigh the costs. This is small comfort to floodplain people. Furthermore, it is highly unlikely that any CBA would have access to information on where people would go or what economic changes occur (e.g. shifts in livelihood or cropping patterns) with which to quantify impacts. There are also costs which are hard to quantify in terms of some kind of economic value, for example the stresses of resettlement or migration.

It is quite possible to try to take these issues into account and carry out a broader social cost/benefit analysis. However, this is often not done and even if attempted may be unsatisfactory as a decision-making tool because of unknowns. The technical planning tool of CBA is only as good as those defining its aims and interpreting its results. Very often CBAs are limited to technical or commercial criteria and ignore wider issues. As a result they are very poor at identifying these unquantified and usually unmeasured costs.

The distribution of costs and benefits of dam projects is often extremely unfair. While it could be argued that decisions about development projects in such cases are essentially political, and hence beyond the scope of technical project appraisal, in practice politicians (and very often bureaucrats in national development agencies such as Ministries or River Basin Development Authorities) have neither the time nor the expertise to assess project appraisal studies thoroughly. If not addressed in consultancy reports, wider issues are often not addressed at all. Usually the problems described here are mentioned somewhere, but when the feasibility reports on a dam form a pile half a metre high, most of it a blur of numbers and diagrams, it is understandable that the small print gets lost. What decision-makers concentrate on is the bottom line, the cost, and where the money is coming from. If the aid donor doesn't blow the whistle on environmental grounds, it is highly unlikely that anyone else will. The project goes ahead, with the environmental impacts a time-bomb waiting to blow. When the project is built and the impacts are felt, planners are usually long gone, and politicians and bureaucrats safely tucked away in the suburbs of African cities.

# CHAPTER SEVEN

# Watering the savanna

*The disappointing performance of many modern irrigation projects can often be traced to their conception as exercises in applied hydraulics on a large scale, rather than as a facility for providing a reliable water input to the farmer (J.R. Rydzewski, 1987)* [1]

## Irrigation: part of the solution or part of the problem?

The image on the screen in the simple conference room was a total contrast to the view from the partly blacked-out window. It was mid-February in Kano State, Nigeria, at a conference on change in rural Hausaland being held at the Bagauda Lake Hotel. It was the fifth month without any significant rain, around the middle of the dry season. Outside, the Harmattan was in full swing, the wind strong from the north and the air thick with dust. Driving down to the meeting, the landscape had been shrouded, the sky a hot lid above the baobab trees, the dry farms and the dust. In the villages of the drylands most of the children had colds, and there was an air of inactivity. All farm work had stopped, and men were either engaged in other activities locally or travelling to trade or find work. Only along the river valleys was dry season fadama irrigation creating patches of green.

On the screen the same world was represented slightly differently. Water gleamed in reservoirs and moved along long canals. Water chuckled over little sluices and trickled over the land surface, sinking in and wetting the soil. Green crop plants sprouted and grew strong, and lines of farmers in clean gowns worked the fields, controlling the water and tending gleaming tractors as they ploughed and harrowed the soil. In the film, time was compressed, for within minutes the same farmers were harvesting golden fields of wheat, riding home on their motorcycles at the days end. Plump men got out of saloon cars to view the harvest, speaking with attentive young field staff in jeans, trainers and T-shirts. Almost everyone was smiling.

The conference was watching a film made by a specialist company to help in irrigation extension work, particularly on the Kano River

Project, a 12,000 ha scheme close by. The film was to be made available for showing in villages to farmers whose land fell within the project. It was intended both to demonstrate the principles of irrigation and how the whole irrigation scheme worked, an important point, since the reservoir was some 20 km away from the irrigated area and large-scale irrigated areas are hard to comprehend from the ground. The film also served two other less immediately obvious purposes. The first, and most significant, was to present to farmers the image of success, to demonstrate how the scheme was supposed to work and so to gain their loyalty and support for the management of the scheme. The film was propaganda, like most extension material, and it was very well done. It was communicating information, but much more importantly it was communicating an attitude and ideology. The message was simple: irrigation works, and it is good to be involved.

The second purpose of the film was to communicate very similar messages to decision-makers within the river basin authority and outside. The images of running water and green fields did indeed provide a dramatic contrast to the aridity of the northern Nigerian dry season. Furthermore, this dry season came at the end of the 1970s, the decade which saw Nigeria, along with much of the rest of the Sahelian and sub-Sahelian belt, suffer a sequence of drought years. Irrigation promised an escape from the continuing spectre of crop failure and famine. Who could not be persuaded by the sheer force of the images of running water and green fields in a dry land? In addition, the film showed the modern production methods of irrigated cropping and indicated their success, and the benefits irrigation could bring to farmer, administrator and (presumably) nation. In the 1970s Nigeria was beginning to earn vast sums from the export of crude oil, and was able to borrow more on the strength of future oil sales. Irrigation schemes looked a good way to capture that money and convert it into long-term wealth-creation.

The ideology of irrigation, which (for want of a neater word) we might call 'irrigationism', has been powerful in Africa. It has been good for international consultancies and engineering contractors, and for the experts who jet out from Northern universities to hold forth on the future of Africa. It has suited young states eager to establish control over remoter regions, and to harness and direct energies within the country to predictable and controlled ends. Keith Hart commented in 1982 that 'the prime need is for large, capital-intensive projects that substantially raise the productivity of a labour force effectively controlled by the state apparatus. Nothing can beat an irrigation scheme for that'.[2] Another commentator wryly noted in 1979 that:

By a curious anomaly the twenty years between 1950 and 1970 which saw so many disappointments in irrigation at field and farm level also saw an increase of 70 per cent in the area of land under irrigation in developing countries, representing very substantial capital investment in new works. For many years irrigation planners and development authorities were not deterred by poor field results in their energetic promotion of irrigation projects.[3]

Yet irrigationism has not simply been driven by the political economy of modern African states. Nor has it been promoted solely out of self-interest and narrowly profit-orientated motives, and while corruption has played a part in the development of irrigation policy in African countries in the last twenty years, such factors alone do not explain the rise of irrigationism.

Over the last three decades, a firm belief in the potential of irrigation has been almost universally held by opinion-leaders in all positions in the international development world, and by those experts at research institutions in the North whose job and vocation it has been to dream and plan for the improvement of the lot of the Third World poor. Decision-makers in Nigerian government bureaucracies in the 1970s, like their counterparts in other African governments and in the aid agencies which advised them, were all schooled to see the potential of irrigation. The landscape of Africa, seen through their educated eyes, was transformed from the landscape outside the Nigerian conference room window into the new and promising world of the irrigation film. Africa was viewed through green-tinted spectacles, and projects were devised to make the reality match the image. Unfortunately, as so often, African reality has refused to be moulded to fit the wishful thinking of outsiders.

## Irrigation in Africa

Arguments about the need for irrigation are simple enough. The World Bank estimates that to achieve food security in Africa, food production will have to grow by 4 per cent per year. It believes that a similar rate of growth in export crops will be necessary to provide foreign exchange.[4] In the last thirty years, average growth in agricultural output has been only 2 per cent, and agricultural exports have declined. So too has the value of the major export crops of cocoa, coffee and cotton ( as well as other exports such as oil and copper). Far from becoming self-sufficient in food, food imports by African countries are rising by 7 per cent per year. Total cereal production rose from

33 million tonnes in 1965 to 44 m tonnes in 1980 and 48 m tonnes in 1987. Nevertheless, cereal imports rose from 4.1 m tonnes in 1974 to 8.1 m tonnes in 1986, of which 38 per cent (3.1 m tonnes) was food aid. Despite this, some 100 million people in Africa are inadequately fed. The problems are not the same everywhere. They are particularly acute in the Sahel, Ethiopia and parts of highland East Africa, and in the semi-arid lands of Mozambique, Lesotho and Angola.

Increased agricultural output can be achieved in two ways, by extending cultivation onto new lands or by intensifying production on existing land. The room for taking in new land is now severely curtailed in Africa. Although in the past Africa has been thought of as a place where agricultural output is limited by labour and not land, shortage of cultivable land is now becoming an acute problem in a number of countries. The FAO believes that the annual rate of expansion of the cultivated area in Africa has only been 0.7 per cent per year over the past two decades. At the same time, some farmland has proved incapable of sustaining continuous cropping, and is of declining quality. What is the solution? According to the World Bank, the answer is to increase productivity on existing land through technological change. This means more intensive use of chemical and organic inputs, new higher-value crops, better integration of livestock and farming systems, better animal and crop husbandry using hand and draught tools, better crop storage – and irrigation.

A recent review of the role of crop breeding and the impact of modern varieties (MVs) on the Third World rural poor finds limited success and limited enthusiasm for plant breeding among African governments. It is not expected that there will be rapid development of new crop varieties suited to African conditions.[5] Similarly, developments in biotechnology are unlikely to lead to major breakthroughs in tropical foodcrops, and if they do they may well not have immediately advantageous impacts on the African rural poor. There are various reasons for this. Biotechnology is focused within the food processing industry (some of which seeks to develop substitutes for African export crops), and even work on crops focuses on cash and not food crops. Furthermore, it is expensive, and most is taking place within large private chemical companies. The products of such research can, in the USA at least, be privately owned, and may well not be accessible to public research institutes, particularly in the Third World, and hence not to farmers. Michael Lipton and Richard Longhurst comment that it is 'unlikely that private firms will steer such services substantially towards small LDC farmers, low-cost food staples, or employment-intensive farm-household systems'.[6]

Nonetheless, Lipton and Longhurst believe that strategies to improve water control, combined with research on improved crop varieties and increased fertilizer use are vital if rural livelihoods are to be maintained (let alone enhanced) in sub-Saharan Africa (SSA). They argue that 'without external water security, e.g. micro-irrigation, farmers may not risk fertilizers; without fertilizers *and* MVs, neither food supply nor rural employment income can often keep up with 3 per cent annual population growth'.[7] Crop breeding needs to develop varieties adapted to marginal and risky environments, but improved water control through some form of irrigation or cropping timetable adjustment will also be required. They argue that 'food security requires water security',[8] and that 'traditional farming systems' (even though ignored by researchers in the past) cannot expand output fast enough to meet demand (3–4 per cent per year for 10–20 years). With strategies (including irrigation) to tackle moisture stress, 'MV-based research is, for SSA's poor, the only real hope. It is the only game in the countryside'.[9]

Most attempts to paint an economic future for Africa emphasize the importance of continued investment in irrigation. The World Bank is among a number of donor institutions which, by the late 1980s, were beginning to be critical of some of the ways in which irrigation had been done in the 1960s and 1970s (through medium and large-scale government-managed schemes), but were still fairly bullish about the potential for private sector developments at different scales (from peasant farmer to capitalist agri-business).

**Table 7.1** The area of irrigation potential in sub-Saharan Africa

| Country | Area irrigation potential (000 ha) | % Developed as % of total (modern *and* small scale/traditional irrigation) |
|---|---|---|
| Angola | 6,700 | <1 |
| Zaire | 4,000 | 1 |
| Zambia | 3,500 | <1 |
| Sudan | 3,300 | 53 |
| Mozambique | 2,400 | 3 |
| Tanzania | 2,300 | 6 |
| Nigeria | 2,000 | 43 |
| Central African Republic | 1,900 | <1 |
| Chad | 1,200 | 4 |
| Madagascar | 1,200 | 80 |

*Source:* FAO Investment Centre, 1986

The problems of obtaining adequate data on the extent of irrigation in Africa were discussed in Chapter 4. The best source is the study by the FAO Investment Centre published in 1986, and discussed in Chapter 4.[10] The FAO report gave an estimate of the 'irrigation potential' in different African countries (Table 7.1). The total potential is thought to be some 33.64 m ha, the largest areas lying in Angola, Zaire, Zambia and the Sudan (Table 7.1). In most African countries, very small proportions of the area deemed potentially irrigable by international 'experts' is actually developed (Table 7.1). Different countries show very different levels of use of this potential resource, ranging from 80 per cent in Madagascar, 53 per cent in Sudan and 43 per cent in Nigeria to less than 1 per cent in Angola, Zambia and the Central African Republic.

In some countries (notably Nigeria and Madagascar), most irrigation is in the small-scale/indigenous sector (see Chapter 4). The FAO estimated that there was 2.64 m ha of 'modern' irrigation in Sub-Saharan Africa, only 53 per cent of the total area irrigated. Of this, 65 per cent (1.74 m ha) was in schemes over 10,000 ha and the rest (0.9 m ha) was in smaller schemes between 500 and 10,000 ha. In terms of land tenure, some 61 per cent (1.6 m ha) was in government-controlled smallholder schemes, 20 per cent was in estates and 20 per cent in private sector irrigation of various kinds (estates or individual holdings).

The countries with the largest area of what the FAO like to call 'modern' irrigation are shown in Table 7.2. By far the largest area is in the Sudan, with 64 per cent of the modern sector irrigation within Sub-Saharan Africa. Ninety five per cent of irrigation in the Sudan falls within this sector, compared to only 6 per cent in Nigeria and 17 per cent in Madagascar, the only two other countries with a total of more than a quarter of a million hectares of irrigation. The other 0.9 m ha of modern irrigation is split between 31 countries, of which very few have more than 10,000 ha (Table 7.2).

**Table 7.2** The area of modern irrigation in sub-Saharan African countries

| Country | Area "modern" irrigation (000 ha) |
|---|---|
| Sudan | 1,700 |
| Madagascar | 160 |
| Zimbabwe | 127 |
| Mali | 100 |
| Ethiopia | 82 |

| Country | Area 'modern' Irrigation (000 ha) |
|---------|-----------------------------------|
| Mozambique | 66 |
| Swaziland | 55 |
| Nigeria | 50 |
| Ivory Coast | 42 |
| Somalia | 40 |
| Senegal | 30 |
| Tanzania | 25 |
| Kenya | 21 |
| Malawi | 16 |
| Niger | 10 |
| Zambia | 10 |

*Source:* FAO Investment Centre, 1986

Clearly, for all the enthusiasm, irrigation development has not got very far in Africa. FAO data suggests that the total irrigated area in sub-Saharan Africa rose by about 148,000 ha a year between 1965 and 1974, and by 157,000 ha between 1974 and 1982. In the vast majority of countries, irrigation is carried out on a very small proportion of the cropped land area (and a far smaller proportion of the total area of each country, including the sometimes substantial areas of uncultivable land). The proportion of cropped land irrigated is highest in Swaziland (34 per cent), followed by Madagascar (27 per cent), Somalia (17 per cent) and Gabon (16 per cent).[11] In the Sudan 15 per cent of cropland is irrigated, in Nigeria 3 per cent, Mali 9 per cent, Mauritania 6 per cent and Senegal 3 per cent. In Kenya and Tanzania only 2 per cent of the cropped area is irrigated. The lack of development is such as to give enthusiastic observers good reason to be optimistic about the future of irrigation in Africa. Indeed, the World Bank (using the FAO's data) identifies a series of countries where irrigation could contribute to expanding agricultural production. This includes a series of Sahelian countries (Chad, Ethiopia, Mali, Mauritania, Senegal and Sudan) as well as Nigeria, Malawi, Uganda and Madagascar.

However, as the World Bank is now well aware, it is one thing to identify the theoretical potential of irrigation as a technical solution to the problems of seasonal arid climates (and to stress the economic potential of irrigated production of export or food crops). It is quite another to make these ideas realities on the ground in Africa. The last twenty years of experience of irrigation in Africa shows only too clearly how far achievement has fallen short of imagination. Ian Carruthers wrote in 1978:

> Good irrigation plans can cover all the trendy elements in the current development literature; being rural, serving income distribution; basic needs, enhancing food surplus; creating employment; and incorporating appropriate technology and sound ecological principles for environmental protection. In the event, most schemes appear to create subsidized income élites; contribute to food production only at high cost; facilitate preconditions for inappropriate mechanization and thus a disappointing employment creation record; and they lead to various aspects of environmental degradation. [12]

These problems are revealed starkly by the record of the attempts to build large-scale irrigation projects in sub-Saharan Africa through the 1970s.

## The large-scale irrigation experiment

Enthusiasm for large-scale irrigation has been widespread within Africa in recent decades. Senegal allocated half its agricultural budget to irrigation in its Fifth Plan (1977–1981), and all the countries involved in the development of the River Senegal through the Diama Barrage and the Manantali Dam planned for massive increases in irrigated area in the late 1970s and early 1980s (Senegal 17,000 ha in the Middle Senegal, Mauritania 21,500 ha). A similar rush towards irrigation took place in East Africa. Kenya had several decades of experience to call upon, because under the African Land Development programme (ALDEV), several medium-sized government irrigation schemes were developed in the 1950s, at Mwea, Ishiarra, Hola and Perkerra. Initially these used the labour of people detained under the Mau Mau Emergency, but they were subsequently permanently settled. The National Irrigation Board was established in 1966, and by 1975 plans were well advanced for a major irrigation scheme on the Tana River. A feasibility study was done in 1975 and after a series of reports, construction of a relatively small area of irrigation (6,700 ha) began on the West bank of the River Tana in 1979 (see Figure 7.1). [13]

   Of all the countries that experimented with large-scale government irrigation in the 1960s and 1970s, it is Nigeria which is probably the most significant, largely because of the remarkable scale of investment in river basin planning that took place. This investment was made

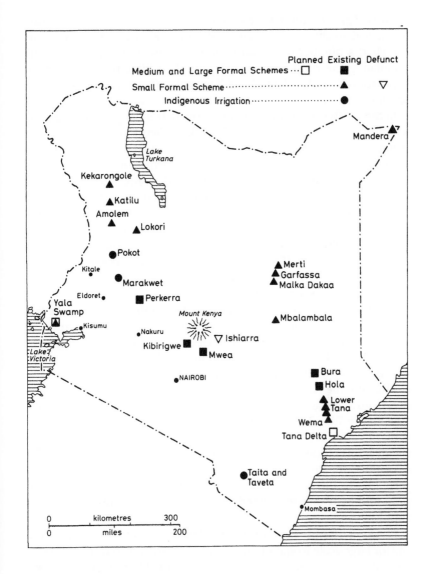

**Figure 7.1**   Irrigation in Kenya

against the background of a depressed and ailing agricultural economy. In 1960, agriculture represented 60 per cent of Nigeria's GDP. By 1980, it had fallen to less than 23 per cent, overtaken in proportional terms by oil and stagnant in absolute terms.[14] Over this period the rural sector atrophied. By the late 1970s, Nigeria had ceased to be self-sufficient in food, and was importing large volumes of wheat and rice.

Exports of groundnuts had ceased, and exports of cotton, palm oil and cocoa had declined. Over the 1960s and 1970s the Nigerian state launched a number of initiatives aimed at re-capturing farmers and stimulating the rural economy. Despite their gung-ho titles, such as 'Operation Feed the Nation' and 'Operation Green Revolution', they had little effect.

In the 1970s massive investments were made in large-scale irrigation schemes in Nigeria, through the River Basin Development Authorities (Chapter 5). Irrigation, particularly to grow wheat to replace imports, was widely seen as a solution to the poor performance of the agricultural economy, as well as the problem of drought. Plans were most ambitious, and optimism was maintained even though the actual area of irrigation grew very little. In 1971, the Federal Department of Agriculture estimated that there would be 320,000 ha of irrigation by 1982; in 1979 the World Bank predicted that 125,000 ha would be irrigated by 1982,[15] and in the 1980–85 Plan period RBDAs planned to develop 2 m ha. In the Fourth Plan irrigation received 2,265m Naira out of the 8,989m Naira allocated to agriculture.[16]

The fruits of this policy were soon visible on the landscape of rural Nigeria. Large headquarters buildings were constructed, and extensive areas of farmland were identified and labelled with imposing signboards as developing irrigation schemes. In the north of the country, a series of formal large-scale irrigation schemes were developed (including the the Kano River Project in Kano State, the Bakolori Project in Sokoto State, and the South Chad Irrigation Project in Borno State) (see Figure 7.2). These projects were characterized by optimism, not caution. Thus the planned extent of SCIP was calculated assuming steady levels of Lake Chad. These levels had been at their highest for a century in the 1960s and subsequently declined continuously through the 1980s, until Lake Chad was reduced to a small southern pool and SCIP was marooned many miles from its proposed water supply.

However, despite the enthusiasm of national planners and aid donors for large-scale irrigation in a number of African countries through the 1960s and 1970s, the actual rate of growth of the irrigated area was not particularly high. In Senegal the Société d'Aménagement et d'Exploitation des Terrains du Delta managed to irrigate only 9,600 ha of rice in the decade 1965–75, and only 200 ha of their 4,500 ha expansion target by 1979. On the Senegal, development of formal irrigation *perimètres* has fallen well behind expectation (to the benefit of flood-recession farmers, who as a result enjoyed a few years of continuing flood release from the Manantali Dam, see Chapter 6). In Nigeria in particular, where investment was huge, the predicted rates of growth seem

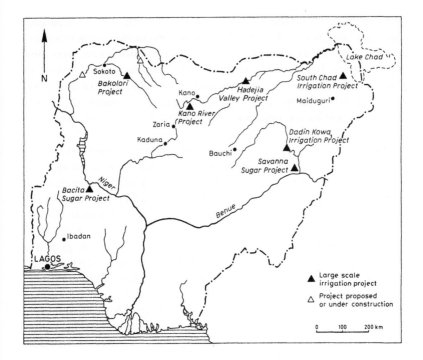

**Figure 7.2** Large-scale irrigation schemes in Nigeria

bizarre when set against achievement. At the end of 1980 less than 31,000 ha were actually irrigated in Nigeria,[17] only a tiny fraction of the area the FAO estimate is irrigated within Nigeria in the small-scale or informal sector (0.8 m ha in 1982, 94 per cent of the total area).

The need to take irrigation development slowly (or, to put it another way, the fact that over-rapid plans are rarely achieved) is recognized by engineers. In a recent text book for irrigation engineers, Clarke and Anderson say that a medium-sized irrigation scheme (5,000–20,000 ha) might take five to ten years from identification to operation. Interestingly, they add a rider to this to suggest that this time can be minimized if there is what they describe as 'a powerful authority responsible for the development'.[18] It might seem desirable from engineering or even narrow financial perspectives to accelerate the planning process. The experience of large scale irrigation development in Africa in the last twenty years would question the wisdom of allowing authorities the power to do this.

Many of the large Nigerian irrigation schemes are far smaller than was originally intended. Thus Phase I of the Kano River Project (KRP) was intended to cover about 16,500 ha. Feasibility studies into irrigation development on the Kano River began in 1969, and a pilot project was begun in 1970. By 1976 the Kano River was being dammed at Tiga and a canal built to supply the project. Construction of the irrigation scheme itself began in November 1977, but it went slower than planned. By 1981 the first irrigated crops were obtained, but only 4,000 ha had been completed. By 1984, 11,500 ha had been finished, but only just over 7,000 ha was cultivated in the 1985/6 dry season. What had gone wrong?

One reason in the Nigerian case was simply lack of funds. The fall in the global oil price in 1983 hit Nigeria very hard, exposing it to a vast volume of debt incurred through the boom years of the later 1970s by stripping away its capacity to pay. This economic crisis strangled the Nigerian irrigation boom. Despite the limited achievements in Phase I, plans for Phase II of the Kano River Project, a further 40,000 ha, went ahead in 1976. A Feasibility Study, outline designs for the major works and detailed design of a trial 2,000 ha had been done by 1981. By 1982 reports on agronomy, resettlement and other issues were complete, and development awaited only financial go-ahead from the Federal Government. To date, construction has not begun, for lack of federal funds. Phase II consists simply of a small 90 ha pilot farm. Instead of the 82,000 ha planned, the whole KRP operation amounts to less than 12,000 cropped hectares.

The story is very much the same elsewhere. Further down the Hadejia River (immediately upstream of the Hadejia-Jama'are wetlands described in Chapters 4 and 6) construction of the first half of a 25,000 ha irrigation scheme, the Hadejia Valley Project, began in 1982. By the end of 1984 much of it was built: a barrage across the Hadejia River, a good length of the main canal and drains and some of the field structures. However, at this point all work stopped because the Federal Government ran out of money. Only in 1991 did new finance from a consortium of French Banks allow work to restart.

However, lack of money was only part of the problem. More serious was the growing understanding by both aid donors and governments of the huge battery of problems with large-scale government irrigation in the African context. A decade ago, Ian Carruthers commented, 'there is growing unease, both among participants and outside observers, of the gap between potential and actual performance in the operation phase of development projects in general and irrigation projects in particular'.[19] Planning, managing and paying for irrigation schemes proved far more problematic than hopeful planners had anticipated.

# Planning irrigation

Those planning developing large-scale irrigation in sub-Saharan Africa have faced a series of technical problems related to environmental conditions. Terrain, soils and hydrology all make irrigation development difficult and expensive in Africa, compared, for example, with South Asia. To these natural constraints may be added a range of socio-economic, political and administrative problems which, while far from unique, are particularly well-developed in Africa. According to the World Bank, investment costs per ha of irrigation in Africa are between 1.4 and 2.4 times as expensive as they are in low-income countries in South Asia (between $6,000 and $10,000 per ha). It argues that this is because of over-valued currencies, difficult terrain, long distances and poor communications, lack of skilled people, poor public sector management (including corruption) and policies which hold back the private captalist entrepreneur.[20]

The problems presented to hydrologists working with data sets covering only short periods under African conditions of great variability in rainfall between years have been described in Chapter 3. They make for a range of difficulties for irrigation planners. The most striking example is undoubtedly that of the South Chad Irrigation Project in northeast Nigeria (SCIP) and the shrinkage of Lake Chad due to drought. This also was discussed in Chapter 3. Extension of the canal in pursuit of the shrinking lake through the early 1980s was extremely costly, and ineffective. The area irrigated has been but a small fraction of that predicted, and of course the economic benefits that were supposed to meet construction costs never materialized.[21] Problems of water supply can also occur in other circumstances, for example with the maintenance of a weir in a river which is undergoing rapid lateral movement, as on the Perkerra Scheme in Kenya. In 1953 it was expected that this would cover 12,000 ha, but in the 1980s only 415 ha were irrigated, partly because of technical problems with the river intake structure.

The limited suitability of African soils for irrigation is also a problem. Not only are many African soils infertile but there can be serious problems of drainage (leading to waterlogging). Poor soils were a major factor in the development of the Bura Project in Kenya. A study in the 1960s suggested that there might be potential for irrigation schemes of 100,000 and 120,000 ha on the lower Tana, once HEP dams had been built higher in the basin. A subsequent study queried this, however, and a feasibility study was eventually done in 1975 on an area of 14,000 ha on the West Bank of the river. This revealed the marginal quality of the soil. A World Bank appraisal report (1977) recommended development of 6,700 ha on the West Bank, consisting of 4,500 ha of suitable soils and 2,100 ha of shallow soils. If this proved successful, development of a second phase of a further 6,000 ha on the West Bank could go ahead, with the possibility of a much larger Stage II project on the East bank and a barrage on the river.

Therefore, the economic appraisal of the Bura Project assumed that more detailed soil survey would support extension of irrigation on both the West and East banks beyond the small 6,700 ha scheme of Phase I. Unfortunately, the soils failed to live up to hopes. The East bank area was never seriously considered for development, Phase II was not built, the barrage was never constructed and the project to this day depends on a 'temporary' pump station (see Figure 7.3). The World Bank's mid-term evaluation of Bura identified the problem of low productivity related to poor soils as a major factor in the project's poor performance (along with poor water supply and slow implementation).

One effect of the rather odd set of decisions taken about Bura was the rapid rise in costs of construction.[22] By July 1979 the rate of return was reduced to 4 per cent, and the World Bank concluded that the project was no longer viable. However, the Kenyan Government was not willing to abandon the project, in part because of what the Bank saw as 'the very strong political motivation to help the landless and unemployed'.[23] The project went ahead under-funded and with rapidly escalating costs. Construction started a year late in 1979. Settlement also began a year late (in 1981), and ran at half the planned rate. The development of irrigation was also therefore slow, and by 1982 Bura was two years behind schedule. Costs were 187 per cent of those predicted in the 1977 report, and (partly because of the withdrawal of one donor) the Kenyan government was now bearing 50 per cent of them. The project wallowed from crisis to crisis, and in January 1986 the Kenyan President criticized it savagely, saying it was 'a failure, a disgrace and a prime example of mismanagement'.[24] It would have been more accurate to describe it as an example of poor planning.

Clearly, the economic appraisal of irrigation schemes at feasibility stage is critical in determining whether they go on to be developed or not.[25] Economists work under difficult conditions, often with very poor data, and are expected to come up with a quantified answer often using very little more than sensible assumptions and guesswork. It is not even clear what crops will be suitable for a new irrigation scheme, either in terms of agronomy or economics. Often, in fact, it has proved that the crops planned for an irrigation scheme are unsuitable. Thus it has proved difficult to find a profitable crop on the Perkerra Scheme in Kenya, and the Mwea Scheme faltered until in 1958 a decision was made to specialize in rice production.[26] At Bakolori in Nigeria,

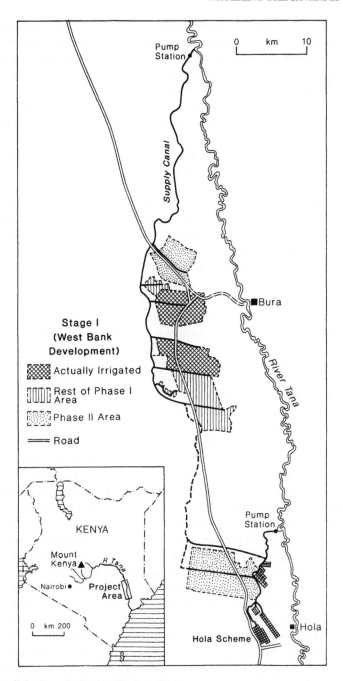

**Figure 7.3** Bura Irrigation Project, Kenya

economists asumed that large areas would be devoted to a sugar estate (which did not materialize because land tenure surveys showed that the large areas of land believed to be unoccupied were in fact claimed by small farmers), irrigated fodder to supply an intensive cattle feedlot (which was never funded and never built), and tomatoes (for a factory which was also not built). Without these innovative ideas, it is unliklely that the Cost/Benefit Analysis would have been favourable enough to allow the project to go ahead. The cropping pattern on which the economic appraisal was done did not, of course, materialize, and nor did the revenues which were confidently predicted.

In a way the calculation of predicted costs and benefits in irrigation planning is rather an academic exercise, in that while aid donors require the economist to show that certain rates of return are possible and a certain ratio of costs to benefits is expected, it is widely recognized that these will be notional figures. Once the real world of cost-overruns, delays, bureaucratic ineptitude and corruption comes to bear on them, and once the quick-look surveys of soils and water resources are replaced with detailed experience during construction, the neat picture created by the economist's analysis is often drastically changed. Nobody is terribly surprised about this, least of all the economist, who probably knew how poor the data used were. Of course, the calculations could easily be re-run with better data at a later stage. But development decisions don't wait. Once the appraisal is done, decisions are made. Then the project juggernaut starts rolling, money is committed, contracts are awarded and work is begun. It is too late then for second thoughts.

One problem with irrigation cost/benefit calculations are the yields predicted for irrigated crops. Where do economists get figures for likely yields under irrigation? From experimental farms or 'pilot' projects. This means that crops grown under careful management of an agronomic experiment, or by selected farmers well supplied with inputs like fertilizer and pesticide (and water) by a numerous, competent and highly-motivated management team are used to predict yields on a working irrigation scheme where none of these conditions are met. It is not surprising that real-world yields are much less than those predicted. On the Bakolori Project in Nigeria, yields ranged from 17 per cent of those predicted for cotton to 56 per cent (for rice) and 69 per cent for groundnuts.[27] While the details differ, the pattern is standard in irrigation schemes around Africa: economists are excessively optimistic about yields. Over-estimation of yields mean that the economics of a proposed irrigation scheme look more favourable than they should. As a result, there is a danger that a scheme will be developed which will not recover its costs.

This tendency is increased where there are inadequate data about yields of the existing crops which the scheme will replace, grown under rainfed or flood-recession conditions. Few economists are in a position to predict yields under indigenous cropping, partly because of a lack of either experimental work by agriculturalists on indigenous crops, or extensive surveys of yields farmers actually obtain. It is not wholly surprising that teams assessing the benefits of large-scale irrigation development assume that yields under indigenous cropping are low, but they often do so on the basis of ignorance. Where they are wrong, their calculations bias project appraisal in favour of formal irrigation over rainfed or indigenous irrigated cropping. Such calculations are a dangerous basis for investing scarce resources.

A related problem in assessment of the benefits of irrigation is the amount of attention paid to other economic uses extinguished by irrigation. Where wetland environments are developed, these impacts can be serious. In particular, pastoralists may lose dry season grazing resources, or may find their access to dry season water supplies or their migration routes blocked. The tribes of northern Kenya, particularly the Turkana and Boran, have been affected in this way by the development of small-scale irrigation projects (ironically, projects developed in part to help settle pastoralists made destitute by drought). The existence of land use conflicts in the river valleys of Nigeria and along the Senegal have been described, and similar problems are widely reported elsewhere. At Bura in Kenya, an ingenious solution was adopted to the problem that the supply canal acted as a barrier to access to the river. Ponds for livestock were built on the side away from the river. Unfortunately, low water levels in the canal have meant that these have been dry.

## Managing irrigation

Uncertainties at planning and design stages feed very directly through to problems of scheme management and economics. A particular and common problem is that of irregular or inadequate water supplies. These have significant economic impacts, both on the returns to an irrigation scheme and on the ability of individual farmers to survive within the scheme. Unpredictable water supplies reduce yields, which reduce project benefits. Low yields are rarely accompanied by a fall in running costs, and the ratio of costs to revenue deteriorates. Project planning is often based on an attempt to impose stability on water supply, often without success. The alternative, management for uncertainty, demands less optimistic forecasts of project performance, but they are likely to be both more achievable and more sustainable.

Lack of reliability in water supply can be caused by a number of factors. Schemes supplied by pumps are particularly vulnerable because water supply is dependent on good maintenance and pump reliability. In Nigeria, pumps stopped at SCIP when the generators burned in 1982, and the Hadejia-Jama'are River Basin Development Authority reported that problems with pumps were causing 'the gradual and continuous decline' of a series of small pump schemes developed in the 1960s on the banks of the River Hadejia.[28] By the mid 1980s only a tiny proportion of the total irrigable area (707 ha) was actually growing crops. The Bura Project in Kenya is also dependent on pumps to lift water into the supply canal from the Tana River. Four pumps were installed, but through the early 1980s suffered a succession of breakdowns due to lack of fuel and spare parts, lack of trained operators and of trained mechanics to carry out maintenance. Water was unavailable for 25 per cent of the growing season in 1983 and 1984. In 1984, the World Bank appraisal suggested that water would only be available for 2500 ha of irrigation. In 1984 the Bank agreed to pay for the installation of another pump, but the difficulties have persisted.

Farmers respond to the unpredictability in water supply by ignoring recommended water use practices, over-watering while water was available, and stealing water by breaking bunds or locked sluices, or bribing technical staff. Such actions by farmers near the source of water ('topenders') mainly serve to make the supply of water to those further down the system ('tailenders') even less predictable. The overall efficiency of water use goes down, and overall yields and economic returns fall. Furthermore such action increases the initial problem of water supply, particularly for tailenders. Tailenders' farms are often smaller and the farmers themselves poorer. In general, the ability of farmers to command sufficient water is closely related to wealth. Poor farmers are much more completely at the mercy of the system, and they suffer disproportionately when the water distribution system is unfair.

Farmers who cannot obtain enough water receive poor yields, yet their production costs (e.g. paying for tractor hire or fertilizer) do not fall. As a result they are severely squeezed financially. Those who cannot make ends meet may leave the scheme, or shift over to crops with which they are familiar and which will survive drought, such as sorghum. These may meet subsistence requirements, but are unlikely to meet the predicted economic returns of the project. They also tend not to fit into irrigated cropping timetables, so that, for example, a farmer growing sorghum in the wet season in a northern Nigerian irrigation scheme will not be ready to grow a dry-season crop like wheat at the time necessary to avoid heat damage. As a result, planting is late, there is a severe bottleneck in demand for ploughing and land preparation services, and dry season yields are also reduced. Thus the poor performance of the scheme is exacerbated, and the economic noose around farmers, particularly poorer farmers, tightens.

One result of over-watering can be waterlogged soils, particularly where drainage systems are inadequate or badly maintained. On the Kano River Project (KRP) in Nigeria, water lay within 1.5 metres of the ground surface before development. Under irrigation, it lay within 60 cm in the wet season and 40 cm when irrigation was taking place.[29] Few plants (rice being the most important exception) can survive in waterlogged soils. In the case of the KRP, it was suggested that improved irrigation efficiency (i.e. ensuring that only the correct amount of water was applied to fields and that losses from canals etc. were minimized) and manipulation of cropping patterns could control the problem of a high water table. However, the only permanent solution would be effective drainage. In many projects (particular those in wetter areas) the design of drains to take water *off* irrigated land is more important than structures to get it *on*.

Waterlogged soils in areas of high evaporation (i.e. hot, dry areas) are also at risk of increases in salinity or alkalinity. Crop plants are highly sensitive to saline or alkaline soils. Half the world's irrigated land is thought to be affected by these problems. Within Africa they are particularly acute in the most arid areas such as Egypt, Sudan and Somalia, but they are significant to some degree in many African irrigation schemes, and where they occur they have a serious effect on yields and economic returns. UNESCO commented in 1977: 'There are numerous examples of soils degraded and lost to production following ill-conceived or poorly implemented irrigation schemes. The most serious difficulties relate not to the detection and delivery of water, but rather to the secondary effects of irrigation, which compromise success in the long run'.[30]

For at least a decade it has been argued that money is better spent trying to rehabilitate existing irrigation schemes, particularly to tackle problems of soil degradation, than to build new ones. This is probably as true in Africa as anywhere else, and in the last ten years considerable sums have been spent in re-designing and rebuilding irrigation schemes, for example on the Jubba River in Somalia. However, there is not a great deal that can be done about waterlogging and salinity/alkalinity without the expenditure of a great deal of effort and money. Engineering solutions include digging new drains, lining canals (to stop water seepage) and even supplementary pumping to lower groundwater levels. Such rehabilitation works are costly, and add to the existing sunk capital load of the project. Work in India suggests that it is the human and organization problems of water distribution and use within the scheme that are most important in determining inefficient water use, and that unless these are tackled costly engineering solutions may not be effective.[31] However, these problems cannot be solved quickly, and require more than a relatively easily-planned (even if expensive) engineering programme. Nobody believes that such problems can be solved easily. Physical rehabilitation is just the start of the task of making irrigation schemes work.

Issues of water supply and distribution are among the most significant management problems on large canal irrigation projects throughout the Third World. Related problems of operation and maintenance ('O and M') include siltation of canals and structures, particularly where water is being drawn from a river draining a semi-arid catchment area with high rates of soil erosion, blockage of canals and structures by floating and rooted plants, leakage of canal lining leading to high transmission loss of water, and the failure of engineering structures. Irrigation schemes require continuous maintenance to deal with such difficulties as they arise. This maintenance requires a properly-funded and efficient works unit with suitable equipment. That means a rolling operating budget to pay salaries for staff, to purchase and maintain earth-moving equipment (much of which will have to be imported) and to buy the fuel to keep it working. 'O and M' thus requires the continued commitment of funds, including foreign currency. This all has to be paid for out of the project's budget. If the scheme is not working well, and costs are not being recovered, this expenditure on 'O and M' becomes a costly recurrent item. It is easy to cut back in this area, but not without further incremental reductions (sometimes of a drastic nature) in the efficiency and economic returns of the scheme.

Management costs on the large scale irrigation schemes tend to be high, in some cases exceeding the gross value of production. High costs come from the costs of water supply (particularly with pumps), the costs of engineering maintenance and the salaries of technical and administrative staff. One example is that of tractor hire services. On the Kano River Project in Nigeria, the river basin authority obtained tractors from the Ministry of Agriculture and from a private company in Kaduna. Neither proved economic. By the end of 1986 the cost of repairs and maintenance was 6.19m Naira, and of running the workshop, 5.6 million Naira. The authority had charged farmers only 0.956m Naira for tractor hire. Like many other elements of large scale irrigation in Nigeria, mechanization costs a great deal of money.

Irrigation schemes tend to suffer considerable crop losses from pests. On average, 42 per cent of crop yields are lost in the field in

Africa, 13 per cent from insects, 13 per cent from disease and 16 per cent from weeds.[32] It is not difficult to see why agricultural development experts see an important role for pesticides of all kinds (particularly insecticides). Even threats once closely-controlled such as migratory locusts became a major problem again in the mid-1980s following closure of the Office International Contre le Criquet Migrateur Africain in 1986 and a hiatus in spraying. Less emotive, but no less destructive, invasions by grasshoppers have also been a widespread problem in the Sahel in the 1980s, and the seed-eating bird *Quelea* is still routinely killed by aerial spraying. Irrigation schemes growing crops beyond the end of the wet season, are particularly vulnerable to such attacks.

However, pesticide use on irrigation schemes has drawbacks. It is relatively costly, and thus increases production costs for farmers. Furthermore, few African countries have domestic production facilities, and so pesticides have to be imported, and paid for using scarce foreign currency. If they are subsidized to promote their use, there is an important exchequer cost which cannot be reclaimed from users. There is also a problem in that it is now known that pests do become resistant to pesticides. The World Health Organization's post-war attempt to eradicate malaria failed because of resistance of mosquitoes to pesticides and malaria parasites to drugs. On the Gezira Scheme in the Sudan problems of pest-resistance emerged with attempts to control whitefly, a pest of cotton. The area sprayed increased from less than one per cent in 1946 to the whole area in 1954. As spraying increased, so did resistance. The response was to spray more often, and by the late 1970s cotton had to be sprayed nine times a year to keep whitefly under control.[33] There are also potentially serious problems of human health associated with the use of pesticides. These are discussed below.

A major factor in the poor economic performance of large-scale irrigation in Africa in the last two decades has been the way in which construction costs proved far greater than estimated. This occurred for various reasons, particularly because of inflation and underestimation of costs at design stage. Underestimation was often due to a failure to foresee problems in feasibility studies. This phenomenon has already been discussed in the context of the Bura Project in Kenya, but it was also a problem elsewhere. At Bakolori for example, the final cost was 400m Naira for 22,000 ha irrigated. This gives a figure of 20,000 Naira per hectare. In their book *The Wheat Trap*, Andrae and Beckman calculate that if the total investment costs in three Nigerian irrigation schemes (the Kano River Project, Bakolori Project and South Chad Irrigation Project) are spread out over a theoretical 50-year life, they amounts to 300 Naira per ha per year.[34] To these must be added costs of maintenance, wages of staff and running costs as well as the costs of the other elements needed to obtain the high yields predicted for the schemes (e.g. subsidized fertilizer and extension services).

Even if no account is taken of the cost of running the administrative structure of the relevant river basin development authorities, Andrae and Beckman argue that irrigation on these schemes would cost about 1,000 Naira per hectare per year. High capital costs of this kind would not matter if economic returns were equally high, but in Africa the returns to large-scale government irrigation schemes have been poor. In practice the high yields and economic returns predicted have usually proved illusory. The northern Nigerian schemes actually require continuous government subsidy simply to meet annual running costs.

## People on irrigation schemes

It is, of course, farmers who end up being asked to pay for the high capital costs and running costs of irrigation schemes. However, threatened by low yields, uncertain water and input supply and costly services, farmer incomes tend to be low and uncertain. Planners often assume that labour is in surplus in African dryland farming areas, and particularly that people are 'under-employed' in the dry season. In fact, of course, most African land use systems suffer critical seasonal shortages of labour. Unless irrigation is designed to avoid (or lessen) these, it is unlikely to be adopted willingly.

Disparaging views of the laziness of peasant farmers appear in the minds even of enlightened planners who seem to think that farmers do nothing in the dry season but sit around in the shade. As a result it is assumed that they will eagerly take up the heavy work of irrigated cropping.[35] In fact there are very often a range of activities through the dry season that have a vital role in household economic survival. The socio-economic impact of low returns from irrigated cropping can be made worse if involvement in irrigation in the dry season prevents people from engaging in other economic activities. These might include labour migration, craft work or movement with livestock. The new activity of irrigation must compete with these established and familiar activities. It must not only replace the income available from them, but it must be seen to do so. Any risks attached to dry-season production (e.g. erratic water supplies) may make irrigation an unattractive alternative to existing activities. Irrigation can also be unattractive for reasons that have nothing to do with economics. Thus among other things it is likely to demand heavy labour in harsh conditions, and it may prevent people from doing activities that are important to them in other ways. Young Hausa men in northern Nigeria, for example, cannot go and take Koranic training if they are tied to an irrigated farm.

Ideas about what 'farmers' do, and the attitudes that they are likely to have towards irrigation can be unrealistic in other ways. Another classic mistake that planners make is to assume that farmers are male. In Africa some 80 per cent of farm work is done by women. Even in Muslim areas such as northern Nigeria substantial numbers of women hold and work farmland. The demands on a woman's time are usually vast, and any assumption about labour availability based on observations about male labour is unlikely to be very relevant. Irrigation development can also create demands for labour for which there is no established precedent in a society. Thus the Jahaly-Pacharr Scheme begun in 1984 in the Gambia, which involves smallholders growing rice under contract, has created new pressures within the household. The labour of women is captured and directed, and their lives made substantially harsher, by the new demands introduced by irrigated cropping.[36] Far from falling on a receptive and under-occupied peasantry, irrigation is deeply coercive, and resembles more the final straw which breaks the camel's back.

Central to the question of the impact of irrigation on farmers and rural households is that of risk. Ironically, although one of the main claims for irrigation is that it provides a guaranteed crop, to the small farmer, life on an irrigation scheme can be extremely risky. The Bakolori Project in Nigeria was developed on land already cultivated in the wet season. Farmers had to stop growing crops while irrigation works were built, and for various reasons it took some time (years in some cases) for land to be re-allocated. During this period farmers fell back on exactly the strategies long used to survive drought periods. The only farmers to do well during this time were those rich enough to set up in business to benefit from the construction work (e.g. running a bush bus service), or who had a member of their household who got work on the scheme itself (e.g. as a labourer or office boy). To others, the introduction of irrigation was an inexplicable stress on their household economy of unknown form and duration: just like a drought.[37] Clearly, schemes where farmers experience problems such as these are unlikely to make the kind of positive impact on rural welfare and income that is usually expected.

Debt is a major factor in the socio-economic impact of irrigated cropping. It can take two forms. First, debt to the project (as farmers are expected to pay for inputs despite low yields), and second, debts to others (for example where farmers have to borrow grain where grain harvest or cash income are too small). Poor farmers easily get caught in a downward spiral of debt, buying grain when it is costly (before harvest) and selling when it is cheap.

One result of this cycle of indebtedness can be the gradual transfer of land from resource-poor to resource-rich farmers and increasing inequality in land and income distribution. On the Kano River Project in Nigeria farm sizes have risen (farms of 10 ha are not uncommon) as larger farmers buy out those less fortunate than themselves.[38] At Bakolori, poorer and politically badly-connected farmers have been unable to obtain adequate land, inputs and services. As a result, small farmers avoid the risks of irrigation by working as labourers on their own land for absentee landowners from the urban élite in Sokoto.[39] This situation was exacerbated by the existence of relatively large areas of un-allocated land on the project following the brutal action of federal police against farmers protesting about unpaid compensation. At least a score of farmers were killed, and many arrested. Many others fled.[40]

Farmers also face problems with produce marketing and paying the charges levied by the project. If they have to market through scheme managers, and their costs are deducted before payment, they may be pushed below the economic margin. At Bura, tenants are paid in January for a cotton crop planted the previous April. The cotton is trucked out to the coast over unmetalled roads for ginning, and tenants are paid the net cost, minus the cost of inputs supplied to them. By 1983, 65 per cent of tenants who had arrived the previous year were in debt. If scheme running costs are too high for the farmers to pay them through charges, the scheme has to be subsidized by continuous government subventions. Most of the large irrigation schemes in Nigeria and Kenya are in this unfortunate position. Farmers marketing their own produce may be subject to low prices due to gluts of perishable produce. There was a bumper grain harvest at the Kano River Project in the 1985 wet season, but lack of outlets meant farmers had difficulty selling grain, and thus had no money available to pay for land preparation, water charges and fertilizer.

Irrigation brings certain very specific problems of public health. In addition to their cost in terms of human misery, poor health and nutrition affect the ability of farmers to work their land and meet the targets of irrigation planners. Poverty among farmers is closely linked to other problems, particularly family health and nutrition. At Bura, a 1985 survey showed that 52 per cent of children on the project suffered from malnutrition.[41] Even at the Mwea Project in Kenya, generally regarded as the best Kenyan scheme, and certainly the only one which does not require continuous subsidization of running costs from central government, there are problems of low incomes and malnutrition among children.[42]

Standing water provides excellent breeding conditions for the mosquitoes that carry malaria, and shallow slow-moving irrigation canals, particularly where they are badly maintained and overgrown with plants, provide conditions for the snail that is the intermediate host of bilharzia or schistosomiasis. It is estimated that 60 per cent of adults and 80 per cent of children on the Gezira Project in the Sudan have bilharzia. Treatment of canals and other water bodies with biocides to kill insects and molluscs would reduce infection rates, but it would be expensive. Both engineering solutions (e.g. making the shorelines of water bodies steep) and management regimes (e.g. the frequency of weed clearance and water drawdown rates) may have a role to play in reducing the incidence of the intermediate host, but neither is easily introduced once a scheme is built and operating. There are also effective drug treatments for bilharzia, assuming the finance and facilities are available, but the problem of re-infection remains. Studies in Zimbabwe stress the importance of minimizing human–water contact, preventing faeces or urine from entering water and reducing the population of snails.[43] Bilharzia is by no means the only disease associated with irrigated areas. There can also be serious problems with more prosaic but just as debilitating intestinal parasite infections. Poor water quality (for example drinking canal water polluted with pesticides) can also represent a problem. Again, children are particularly at risk.

A public health and medical treatment programme ought to be part of planning for any large-scale irrigation project, but this is not always the case. At the Bura Project in Kenya a Health Centre was built by 1981, the year in which settlement began, but it was not opened until 1983 because the Ministry of Health lacked funds to commission it, and was still only partly operational in 1984. At that time there were 1,800 settler families at Bura, from all parts of Kenya. Disease was seen as a major problem, and many settlers sent their families back to their home areas (thus losing the 'free' family labour that the planners assumed would be available to pick cotton). Unsurprisingly, desertion rates were high.

There is a further threat to human health associated with the levels of pesticide use necessary to gain the high yields required of irrigated agriculture, particularly where farmers are unable to read safety instructions. Every year there are about 375,000 cases of pesticide poisoning in the Third World, and 10,000 human deaths.[44] Organophosphorus pesticides such as endosulfan are contact poisons, and it is vital that they do not touch the skin. For obvious reasons, few African farmers have access to suitable protective clothing. Very often they may be unaware of specific safety needs of the chemicals they use. Organochlorine pesticides, such as DDT and Dieldrin, are cheap, and relatively safe to human users; however, they have mostly been banned in industrialized countries because of their persistence in the environment. They are widely used in agriculture in Africa on irrigation schemes and elsewhere, and drain into water courses used for water supply by downstream communities. The extent and long-term implications of pesticide use in Africa are unknown.

## Small-scale irrigation

The poor performance of large-scale irrigation in Africa is now widely acknowledged. Like many people whose worldview was influenced in the 1970s by Schumacher's book *Small is Beautiful*,[45] my first reaction to the waste and foolishness of large-scale irrigation in Africa was to blame its failure on its scale. It stood to reason that projects which were too large would not work. Small-scale irrigation stood a much better chance. There seemed to be ample support for this view in the patchy history of development projects in rural Africa. To an extent, I still hold this view, but as time has gone on I have come to modify my initial simplistic thinking. The simple truth is that there are large numbers of small-scale irrigation schemes in Africa that work no better than their more infamous larger neighbours.

A number of donor agencies have been attracted to small-scale irrigation, partly as a result of the poor performance and high costs of large-scale projects. This interest has been shared right across the spectrum of aid donors, from multilateral agencies such as FAO, the World Bank and the United Nations Development Programme through a whole range of bilateral (national) aid donors, including USAID, to the small 'non-governmental' agencies such as OXFAM, Band Aid and church groups. Small projects commend themselves for many reasons, perhaps chiefly because they promise greater cost-effectiveness. A report by UNESCO in 1977 on the development of arid and semi-arid lands criticized the bias towards large-scale agricultural projects, saying that 'modest and relatively less costly projects would have a greater impact on the economy and the improvement of the lot of the people'.[46]

In practice, small-scale irrigation schemes have also been rather costly, and bad at meeting the needs of the poor. Very often they have been little more than scaled-down versions of large projects, with similar high-technology irrigation systems developed through a top-down planning process and administered by a government bureaucracy. This is certainly true of the small-scale irrigation projects developed in Turkana and Isiolo in northern Kenya over the 1960s and 1970s.[47] The Turkana projects started from an unpromising base, as part of long-term famine relief, from the 1960 onwards, developed by a range of NGOs. Attempts to use wind and donkey pumps failed, so small dams and structures were built to divert water from the Kerio and Turkwel Rivers. The projects were small (up to 125 ha), and Turkana pastoralists were settled on them. However, although small, the projects involved mechanization and central control of water allocation and cropping. At different schemes there were serious problems of water shortage, flooding and salinity. At Malka Dakaa in Isiolo, there were similar problems caused by the erratic regime of the river, soil salinity and the remoteness of the project's location.

Despite this uncertain start, FAO and UNDP invested in the expansion of irrigation in Turkana in 1979. The result was disastrous. Only a fifth of the funds promised by the Kenyan government for local costs were forthcoming, and the area irrigated actually fell between 1979 and 1982. Mechanization cost 1.3 times the gross value of crops produced on the projects, water supply was erratic, yields were low, as were settler incomes. The 700 Turkana settled on the projects were in a poorer state nutritionally than those who had re-established themselves as pastoralists, and the projects had cost over $63,000 per ha (1983 figures), or over $0.4 million per irrigating household (equivalent to famine relief for 200 years).[48] Meanwhile, livestock owned by Turkana settlers was causing local overgrazing around the projects, and fuelwood demands were causing degradation of the riparian forest which formed an important dry season forage resource for Turkana still herding livestock. Indeed, there is evidence that the irrigation schemes themselves were seen by some Turkana merely as one more resource that a household could call on, a source of cash income and an insurance against drought.

The problems of these schemes are not hard to enumerate. The development strategy for Turkana District for 1985–6 saw little future for the experiment.[49] The FAO projects suffered from over-design, excessive reliance on mechanization, high capital costs and top-heavy bureaucratic management. The NGO schemes had a high degree of outside control, and depended so closely on the inputs of individual dynamic leaders from outside that they tended to collapse when this person left. All the projects suffered from periodic problems of too much and too little water (i.e. drought and flood), problems with weeds, and economic problems associated with their remoteness.

These problems are fairly typical of those experienced by small-scale irrigation schemes developed by African governments elsewhere. Economists point to the problem of diseconomies of scale at each stage of the project development process. At the planning stage, separate technical reports need to be done on each scheme. The problems of lack of data and the difficulty of predicting environmental change can be as severe on a small project as on a large one, and a serious result of this is the possibility of uncertain water supply. Furthermore, each separate small project needs its own water supply system (weir, small barrage etc.) and the resources (equipment and staff) to maintain it. During construction, there are costs associated with remote location and small-scale construction contracts. Per-hectare construction costs tend to be up to a third lower than for large-scale schemes, partly because some of them devolve onto farmers, although this is not always the case. Costs of small water-source development can vary widely, both in terms of capital cost and running costs: shallow tubewells, for example, are relatively cheap to sink, but the pumps necessary to operate them can be expensive to run and maintain.

Once built, there can be difficulties in servicing small-scale projects, particularly if they are in remote locations. Small pump schemes along the River Niger in Mali (18–109 ha in area) rely on competition between northern aid donors to supply funds to pay for depreciation and repair and to produce supplies of diesel and fertilizer. The pump technology is not as a result in any sense self-sustaining.[50] Supplies of extension advice, seeds, fertilizer and pesticide are difficult to provide in adequate quantities and at the time they are needed, and as a result yields can be low and variable. There can also be problems with the marketing of produce from small-scale schemes. A study of a small (15 ha) pump-supplied communal garden project in Gambia found similar problems. The scheme, run cooperatively by women, involved a diesel generator to produce power for two electric pumps, and irrigation by sprinkler and spray irrigation. High running costs were threatening the economic viability of the project within two years of establishment. Problems included the costs of fuel and maintenance, of seeds, fertilizers, tools, materials and the wages of a watchman.[51]

It is also a factor that a large number of small-scale projects is needed to match the area irrigated in a single large scale project. Where skilled manpower is scarce, and where administration is costly (because of bad communications, for example) and inefficient (because of poor training and support), such a programme can be problematic. In this circumstance, a single large-scale scheme could look very attractive both to donors (anxious not to see their money dissipated) and national government (eager to retain control). Thus, it might seem attractive to build a single 100,000 ha irrigation scheme rather than 1,000 hundred-hectare schemes. If the bureaucratic infrastructure and competence are weak (as they mostly are in Africa), neither is likely to be very effective. Whatever the scale of development, it looks very much as if it is the way it is run which is of the greatest significance in terms of success. Centralized bureaucratic control by government agencies rarely brings success.[52]

Small-scale projects can also have impacts on local economy and society that are both unplanned and far from trivial. Schemes have been introduced along the middle Senegal River in Senegal (by SAED) and in Mauritania (by SONANDER).[53] During the first phase of this development (1975–1985) most schemes were located on higher sandy areas within the floodplain, and thus favoured those groups of people with land in this area. Villages of fishermen, for example, were strategically located, and as a result have an unusually high proportion of schemes compared to those of *décrue* flood-recession farming specialists. Irrigation development has caused problems for Fulani who seek to move stock into the floodplain in the dry season only to find their access blocked.

There are also socio-economic impacts within villages. Land allocation on the new schemes is on the basis of permanent residence, with one plot of fixed size being allocated per head regardless of wealth or status. This favours the poorest and lowest-status groups (e.g. freed slaves) over the established élites, many of whom have migrated to cities (or to France) on a quasi-permanent basis. This egalitarian policy has in some cases had rather a revolutionary effect. However, there is now a danger that irrigable land will be acquired by those groups traditionally powerful in the valley, and by urban entrepreneurs, under laws allowing statutory tenure to be established. Whatever happens there is no doubt that the symbiosis that once held between wetland and dryland cultivation, fishing and pastoralism has been shattered by the arrival of formal irrigation.

Furthermore, irrigation in the Senegal Valley is not achieving its hoped-for efficiency. Irrigation infrastructure has deteriorated, largely because of the policy of '*le Désengagement*', or withdrawal of state support in response to conditions imposed as part of IMF structural adjustment loans.[54] Mechanization has also been very slow, and the extension of irrigation is therefore causing significant seasonal shortages of labour. As a result the double-cropping of rice planned for in the village schemes has not taken place, and yields have been low due to delayed planting or planting without inputs. Rice production has therefore remained static despite an expansion of irrigated area.

At the same time, the development of irrigation schemes has become the principal method for acquiring land, and there is sustained pressure for the expansion of the irrigated area not just from those seeking to increase output from the irrigated sector, but also those who see this as a means of acquiring land. Competion for land has sharpened with government support for outside entrepreneurs, and this has become a significant factor in the growing conflict over land in the valley. One commentator argues that 'what is likely to happen in the future is that irrigated tracts will be owned by wealthy people, either successful migrants or powerful and affluent strangers, and parcels will be leased out on agricultural tenancies'.[55] Riverside irrigation in Senegal is creating 'a rural labouring class without rights to land'. While others regard the Senegal schemes more favourably,[56] there seems good reason to question both their economic effectiveness and their social impact. The logic of a development strategy based on the disruption of established flood-cropping practices in favour of such irrigation, using the Manantali Dam to control the annual flood, is obviously also highly questionable.

## Agribusiness: the future of irrigation?

Formal government irrigation has won for itself a deservedly poor press in Africa. Most attempts to transform Africa's environment and agriculture through formal irrigation have failed. Reviews highlighting the key problems of large-scale irrigation (high cost, poor planning and management, excessive technical complexity, lack of expertize and lack of basic research, the shortcomings of African bureaucracies and the general 'top-down' approach to planning)[57] could apply just as aptly to small-scale irrigation. Aid donors have not been blind to this, indeed it is research that they have funded that has chiefly served to highlight the problems. While the message about the failure of irrigation still has a long way to go before it has permeated the depths of engineering companies and development bureaucrats, the opinion-leaders have undergone a major attitude shift. What is their response?

Undoubtedly there has been a slow-down in the rate at which new irrigation projects are being commissioned. However, this does not mean that donors and national governments are turning away from irrigation; far from it. Irrigation remains a key element in strategic thinking about the future of rural Africa. The ideology of irrigationism also still has adherents, both within national governments and the donor and consultancy communities. There are those who believe that the environmental impacts of dam construction can best be dealt with by building more dams and shifting water from one river basin to another, or by building canals around swamps to cut 'losses' of water. Similarly, although large-scale irrigation has not worked in the past, there is an argument that it will work given the right technologies (perhaps sprinklers not surface irrigation) and if organized properly (by controlling farmers so that they don't misuse water and do grow the right crops). In the minds of planners large-scale irrigation is still a going concern in Africa. Despite the bad press, there are still engineers to design it, governments to buy it and aid donors willing to lend the money for it.

New approaches to irrigation development are being considered by donor agencies as the failures of large and small-scale smallholder irrigation sink in. One option being canvassed is to continue with large-scale irrigation, but to move from government schemes with large numbers of smallholder farmers to large capitalist farming operations – private farms and estates, including those owned or managed by transnational agribusiness companies. Such irrigation can have a very direct link into world markets and First World consumers. In winter, fruit, vegetables and salads are brought in from surprising places to the fresh produce shelves of British supermarkets. Recently, crisp speciality vegetables have been appearing in neat plastic boxes with a little aeroplane on the label, flown in from a number of places in the Third World, including Africa, where they are grown under irrigation as a new export crop.

Kenya is one source of these vegetables, particularly green beans. Medium to large-scale commercial farms comprised about 20 per cent of the total irrigated area in 1982, and the area has certainly grown since then.[58] The cultivation of irrigated vegetables for Europe has expanded rapidly in the last 5–10 years. About 90 per cent of coffee on holdings of more than 50 ha is also irrigated, and there is an important trade in irrigated flowers, flown to Amsterdam where they enter the world market. In September 1990 the same vegetables were widely available on the street corners of downtown Nairobi: the air freight companies had raised their prices, and there was a stand-off. The supply lines from rural Kenya to the supermarkets of northern Europe are long and have their own hazards.

The use of water in Africa to irrigate high-value crops for Europe may seem bizarre, and in the context of a country such as Kenya with substantial levels of malnutrition, perhaps distasteful. Nonetheless, it makes economic sense of a sort. It fits well into the World Bank's call for growth in the agricultural sector in African countries, and specifically for growth in export crops. The argument is that Africa should identify and exploit its natural comparative advantage, where it has one. Declining prices for traditional agricultural commodities (cocoa, cotton, edible oils, tea and coffee) and the ability of Northern industries to create synthetic substitutes for these products demand that Africa looks for new crops with which it can earn foreign exchange on the world market. Out of season green vegetables are one of the 'success stories' so far.

Medium- to large-scale private irrigation requires relatively little adaptation in established patterns of thinking about and working with irrigation in development agencies and institutions. Skills and methods, as well as vested interests, developed through large-scale water resource projects can continue, only now they will be directed not at a large population of smallholders and an inefficient government bureaucracy, but at a streamlined 'modern' enterprise with all the financial efficiency and sophisticated technology of industrialized agriculture in the North. The World Bank argues that 'a vigorous private sector could process and market agricultural produce efficiently, and rising investment could combine with new agricultural technology to steadily raise yields'.[59] Large-scale agribusiness that can integrate agricultural production, marketing, and processing is more likely to develop new products and markets through 'better quality of produce, more aggressive marketing, and less government interference'.[60] What is needed is a more open policy that attracts domestic and foreign capital and technical and marketing expertize: 'Educated Africans who might spurn peasant agriculture could be attracted to work in such modern agricultural enterprises'.[61]

Not only does this private large-scale irrigation strategy fit the World Bank's conception of Africa's economic future, it also fits the conservative ideology that has dominated many Northern industrialized economies in the 1980s: roll back the state, facilitate and reward enterprise, let the entrepreneur loose and jobs, wealth and a viable economy will be created. Some bilateral donors find the shared ground between this political ideology and the lure of irrigation management that will (finally) be technically and economically efficient, irresistible. It panders to the deep-seated prejudice about the incorrigibly inefficient and corrupt nature of African governments, and also to the long-established vision of a 'modern' and high-technology solution to Africa's development problems: an inoculation of Northern industrialized methods that will start to heal the crippled economies of African states.

Benefits to the rural poor will not come through direct increases in food availability but through closer involvement in capitalist enterprise and market exchange. Larger agribusiness enterprises can sign production contracts with smallholders, 'who will thereby receive the benefits of modern technologies, quality control, marketing and other services'.[62] Rural people will also get jobs as labourers on the projects, and will have access through wages to a cash income. They will be removed from direct exposure to the risks of the natural environment of Africa, but will be exposed to new risks associated with fluctuations in the international market (particularly for luxury foods) and the profitability of business enterprises.

Agribusiness irrigation projects can of course help the national balance of payments in African countries, earning foreign exchange that can then be devoted to other welfare projects (health or education, for example). However, production of the base materials for the international food industry involves the lowest gains in value. Without processing facilities the major profits of the international food trade, and all the industrial trickle-down effects that they might bring, will be outside Africa. Even those profits that can be captured by weak and inefficient state bureaucracies are unlikely to reach the rural poor in any useful way. Money will be needed to build and maintain specialized infrastructure to support the export trade (e.g. development of international airport facilities). In most African countries the money gained will be dwarfed by the vast expenditures on weapons, the military establishment and the prosecution of wars.

The promotion of large-scale private irrigation can not only be aimed at export crops but also at locally-consumed food. Such a foodgrains supply policy is biased in favour of urban consumers rather than rural producers. Higher production and lower foodgrains prices squeeze small producers for whom small-scale sales are an important source of income. Ironically, increased grain production based on capitalized industrialized production could exacerbate the risks borne by the rural poor.

The World Bank is eager to foster medium and large-scale enterprise in agriculture (and industry for that matter), but its stated policies also recognize the importance of the small-scale sector and the primacy of tackling the plight of the rural poor through projects that are sustainable and equitable. The Bank argues that even small farmers need to be released from the stultifying grip of the state on inputs supply, produce marketing and economic policy. 'Progress is more likely to be made if farmers are put in control – and allowed to market freely; to invest freely; to establish their own cooperative credit, input supply, and marketing enterprises; to manage their own irrigation facilities; to own the land they work; and to take responsibility for protecting the environment'.[63] There is an important line, however, between rolling back the state and its burdens on the poor and rolling it back so far that they are exposed to the tight and hazardous grip of agribusiness. When

the state withdraws, who holds the ring, and who guards the interests
of the poor?

# CHAPTER EIGHT

# Water, people and planning

*dutse ba ya zama ruwa*
*(you can't get water from a stone)[1]*

## Wetlands and water resource development

The idea that water resource development should involve the construction of large dams (Chapter 6) and large-scale government irrigation schemes (Chapter 7) is less universal than it was hitherto in Africa. It has been challenged by environmentalists and those concerned about impacts on the rural poor in floodplain areas, and the massive costs and poor economic record of such projects have alarmed economists in African governments and aid donors. Nonetheless, the dominant ideology of wetland development continues to stress the need for control of natural processes and the need to transform economic activities. There is still confidence in Northern science and technology, and widespread assent within the development community that the interests of rural Africans are best served by their integration into national and international cultures and economies. To many people, development still means transformation and modernization.

It will be clear from the arguments in this book that I believe that there is a great deal amiss with this strategy. There are serious limits to the adequacy of technical understanding of the African environment. There is a consistent failure to understand existing resource use systems, particularly the way in which they are inter-linked, and the planning procedures used have proved themselves again and again incapable of coming up with projects that do not waste money, damage the environment or impoverish the rural poor. The ideology of wetland development is powerful and self-confirming. It is also very destructive and very wasteful.

Dam construction and irrigation in Africa have rarely to date produced development that is sustainable. The challenge now is to seek alternative strategies for action which have some better prospect of success. The chance of starting from scratch has gone. Much of the success. The chance of starting from scratch has gone. Much of the length of the major African rivers is already controlled by dams, and a great many countries have not only an irrigated sector, but also plans (and finance) to develop it. Dam construction and irrigation are facts of life in Africa. How can they be pursued in ways that promise escape from the disasters of the past? Changes in conventional development strategies will require innovative and holistic thinking and also institutional change: new ideas, and also new ways of putting them into action. The purpose of this chapter is to start to explore where these changes can be made.

## Intervening in indigenous irrigation

In the hills on the West side of Lake Natron in northern Tanzania, the Sonjo people practise hill furrow irrigation.[2] There are about 18,000 Sonjo, living in seven villages. They live in a remote area, and access is difficult. Most of the villages are located half a day by Land Rover from the District Headquarters at Loliondo North towards the Kenyan border up a rocky escarpment road. There are, however, currently no working vehicles in the valley. The area is dry, with only 4–600 mm rainfall, falling in two rainy seasons. The Sonjo practise rainfed agriculture, but also irrigate. Some villages make use of water flowing from a series of springs in the foot of the escarpment above the valley in which they lie. Others divert water from a sizeable river using brushwood dams up to 3 m high. In some villages concrete has been used in some of the headworks which divert water from stream to irrigation canal, but the canals themselves are simple structures cut into the ground or carried over streams on simple log bridges. Stone does not appear to be used. The use of water for irrigation by the Sonjo is simple in the sense that it is released from earthen canals to flood across the surface of the fields. However, water rights are closely defined (and closely linked to social structure). There is a clear cycle of water use and system of water allocation. As a result, water use is carefully managed.

Three kilometres east of one village, Kisangiro, there is a new innovation in the Sonjo irrigation system. In one place, water from a spring has to be carried over a broad shallow stream (about 30 metres wide) to irrigate land on the far bank. In the past, hollow logs were balanced on wooden trestles to provide a simple but leaky aqueduct. Now the river is spanned by a modern structure, with a square-section channel built of wooden planks on sturdy wooden legs built on firm masonry foundations. This was built by engineers from a non-governmental organization, and it forms part of a new irrigation policy in Tanzania, the 'rehabilitation' of indigenous irrigation systems.

Tanzania is estimated by the FAO to have 115,000 ha of small-scale/traditional irrigation (and only 25,000 ha of modern/large-scale irrigation). Areas such as Kilimanjaro Region have substantial areas irrigated by small-scale furrow systems (38,000 ha), notably by the Chagga on the slopes of Mount Kilimanjaro. The extent of this irrigation has led to considerable interest by government and aid donors, and the establishment of formal projects to 'rehabilitate' indigenous systems. A small unit was established with UNDP/FAO funding in the Kilimanjaro Zonal Irrigation Office in 1987 with the aim of improving water supplies, water distribution and irrigation practices and where necessary combating problems of waterlogging and salinization. There are other such projects in Tanzania, for example one by the Dutch Volunteer Agency (SNV) in the Pare Hills, and there is similar interest in Kenya. A Dutch aid project in the District of West Pokot in the northern Rift of Kenya involves small-scale work on furrow irrigation systems, and in Marakwet the Kerio Valley Development Authority has teams working with hill furrow irrigators to repair and supposedly improve their irrigation systems.

Elsewhere in Africa there are innovative examples of development agencies (often non-governmental organizations) working to expand or enhance the productivity of existing irrigation practices. These include, for example, the interest in rainwater harvesting techniques in northern Kenya, in the use of small terraces for runoff capture and erosion control in Burkina Faso, or the attempt to develop partial water control in the 'fadama rehabilitation' projects in the floodplains of northern Nigeria or the Molapo Development Project in the Okavango Delta in Botswana.[3] In the floodplain of the Hadejia River in Nigeria, the Kano State Agricultural Development Project has used simple engineering survey to identify closed basins whose flooding can be controlled by simple concrete and wooden sluices. This initiative, 'fadama rehabilitation', is intended to pave the way for the introduction of new higher-yielding rice varieties and associated production methods (fertilizer, pesticide and transplanting). The idea is that the timing, depth and duration of flooding can be controlled to maximize crop production. The approach is now being expanded in adjacent Borno State under the European Community North East Arid Zone Project. Development based on simple flood-irrigation techniques is advocated elsewhere, for example on the Niger in northern Mali. As Ton and de Jong comment, 'it is worth building on existing flood irrigation technology because it is highly adapted to the prevailing social and economic conditions'.[4]

In other instances, new technologies have been introduced in an attempt to enhance or develop indigenous production, for example small petrol pumps or solar pumps. The introduction of small Japanese petrol pumps in the early 1980s had a remarkable effect on residual soil moisture farming and small-scale shadoof irrigation in the flood-plains of northern Nigeria.[5] Pumps and shallow tubewells were first introduced as part of the Integrated Agricultural Development Projects (part funded by the World Bank), but within a few years their use had spread to other areas. In the Hadejia-Jama'are floodplain in the late 1980s, for example, 56 per cent of the households irrigating in the dry season used petrol pumps. Only 10 per cent used shadoofs or a simple calabash bucket. The rest farmed using residual soil moisture only.[6]

Farmers with access to suitable floodplain land, and the capital or credit necessary to purchase pumps and if necessary tubewells, have achieved major expansion of dry-season vegetable production, and are supplying urban consumers throughout Nigeria. Furthermore, in 1989 the Federal Government banned the import of wheat, and this has generated a pump-irrigated 'wheat boom' in areas such as the Hadejia-Jama'are Valley. Solar pumps for irrigation can be competitive in cost with petrol where farms and the height over which water has to be lifted are small (e.g. less than 2 ha and a lift of 5–10 m). Windpumps can also be effective, but are sensitive to the variations and unpredictability of windspeed (something on which planners rarely have data).[7]

A number of aid agencies have expressed interest in small-scale and indigenous irrigation, and the FAO's data on the extent of irrigation within this sector has added considerable fuel to the fire of this interest.[8] The FAO estimates that the 'expansion' of traditional/small scale irrigation projects may accelerate to 140,000 ha per year by the end of the century.[9] Such ideas form part of a new wave of thinking about development that has taken on board the critiques of the old large-scale top-down development, and has responded intelligently to it. These projects are cheap to finance, they use simple technologies and work at the community scale. Action can (as in the case of the Sonjo aqueduct) be generated by local requests. New technologies (such as the Nigerian pumps) can 'take off', with spontaneous adoption well beyond the reach of the initial project promotion. These kinds of interventions into indigenous irrigation seem the perfect outworking of the idea that 'small is beautiful'. If large is bad and small is good, surely these projects will work? In practice it sometimes turns out not to be so simple.

There is little experience yet with the impacts of projects like these in Africa, but studies in Asia suggest that there are a number of problems with government intervention in irrigation systems, even at this community scale. First, intervention can alter the way irrigators see their irrigation activity. Work on the rehabilitation of tank irrigation systems in Sri Lanka shows that the investment of government resources changes the attitudes of farmers to both their rights and their duties.[10] What was once their irrigation scheme becomes, in some sense, government property. Whereas once farmers saw maintenance as their own responsibility, now they feel it should be done by the government. As a result, upgrading and rehabilitation projects can stultify instead of regenerating indigenous irrigation.

Second, there is a large element of uncertainty about the impact of government intervention in small irrigation schemes. A recent review of irrigation design sees it as a 'single loop' learning process, wherein engineers make selective use of survey data and take action. The result can be unexpected, and can be adverse. In practice, such interventions on existing farmer-managed irrigation systems 'promoted a series of dynamic and partly unpredictable responses'.[11] What this means is that irrigation design is not the clean and predictable process that some engineers would like to believe, but a somewhat chaotic one whose outputs are not entirely predictable. This is obvious on large-scale projects, but is in fact just as important with small interventions in indigenous irrigation. Just as large-scale projects often turn out to be very different from their blueprints, small designs too bring with them a large amount of guesswork. Designs for interventions in indigenous irrigation may not cost much money, but (unlike many large-scale schemes) they run the risk of damaging something that is working fairly well to start with. The uncertainty is greatly increased where interventions are made without adequate understanding of the systems already in place.

Third, the technologies used in strengthening or rebuilding irrigation systems can themselves create risk by fostering dependence on outside support. The simple concrete and corrugated iron used to repair the hill furrows on East Africa is certainly low-technology when compared to the technologies used in the large-scale dam and irrigation projects built elsewhere in Africa, but they are not necessarily materials that farmers in irrigating villages can obtain without external agents. Similarly, the sustainability of pump irrigation depends on the continued availability of spare parts (and mechanics). Once repair could be done swiftly by communal labour using locally-available resources. This may no longer be possible. Repair and maintenance work now depends on the provision of resources from outside. And who provides these resources? A government irrigation bureaucracy: just the kind of agency whose inefficiencies are a factor in the poor performance of large-scale projects. Unless bureaucratic problems of communications, cashflow for procurement, baseline funding for transport and staff can be overcome, there is a danger that 'rehabilitation' projects will expose irrigators to new risks through dependence on outside intervention that they cannot control.

New technologies can transform indigenous irrigation systems, and can have significant and unexpected impacts, particularly on economic and social relations. Even the spontaneously-adopted and obviously fairly appropriate technology such as the Nigerian pumps carries associated risks. There are technological problems looming with this technology, not least a shortage of spares. There are also economic problems of produce gluts and lack of access to markets. There is a serious rise in inequality associated with land-grabbing by large landowners (often urban businessmen), and conflicts between farmers and herders over access to floodplain land in the dry season.[12] There are also problems of excessive drawdown of shallow and confined aquifers. None of these problems were foreseen in Nigeria. The fact that the small pump technology was cheap, appropriate and popular did not mean that it was free of potentially serious drawbacks.

Clearly the apparent runaway success of small scale development does not reduce the need for integrated resource management. Above all, this is necessary to handle conflict. The 'fadama rehabilitation' approach used in Nigeria has stimulated conflict at local level. Farmers with land flooded to different depths have different ideas on the way the flood-control sluices should be operated. Fulani herdsmen, arriving as usual in search of pasture as the flood recedes, find crops growing still on land that was previously available to them. Back in 1920 a very similar approach was used in the Sokoto Valley in northwest Nigeria, building bunds to increase the area of rice production. This ran into very similar problems of conflicting interests and unpredictable flooding, and it was rapidly abandoned.[13] It is ironic (although of itself no bad thing) that after half a century of experimentation with larger and vastly more complex and costly approaches to irrigation, simple ideas are reappearing among those who are trying to develop Nigeria's water resources. However, it is clear that the fact that a project is small in scale, involves simple technology and is cheap gives no guarantee of success. It is easy to intervene in indigenous water and land management systems. It is much harder to do so responsibly, and to be able to predict the likely results. There is an ongoing task of helping those affected to cope with (and benefit from) the changes that result.

The notion that indigenous systems work if you only let them alone is attractive, but it ignores realities on the ground. Indigenous economies and societies are under pressure from outside intervention and influences of all kinds. These pressures cannot simply be switched off. For better or worse, Africa is now locked into the world economy and its accompanying cultures. Indigenous irrigation systems are products of change over many centuries, from slave-trading to the paternal neglect of colonialism, the penetration of money and the grip of the consumerized market economy, the impact of western-style education and the whole panoply of cultural impacts projected by state and world economy in the name of modernization and integration.

The pressures on the sustainability of these systems are very real. Rural communities may well not be aware of the significance of changes in environment or political economy, and may certainly be unable to counteract them. People, particularly the poorest people, in rural Africa often need help from outsiders. It would be foolish to suggest that Africa has no need of the investment, technology, support and even justice that development aid can bring. However, this does not mean that there is a simple 'technofix' solution to Africa's problems that outsiders can introduce, that can be replicated and that will work. To think this is to make the same arrogant error that has strewn Africa with the remains of failed water development projects in the past.

Development problems are not simple but multi-faceted. They are not easily solved, but are inextricably tied up together. They are very often not technical, but social, economic or political. A project that goes in to tackle narrowly technical 'problems' with engineering design and intervention, or which confidently predicts that problems such as water distribution or water rights can easily be changed for the better is unlikely to be very different in its impacts to the many unsuccessful irrigation and water resource projects that preceded it.

Intervention in indigenous irrigation and floodplain production systems needs to be turned on its head. The words being used by those developing small-scale community-level projects are right enough. It is the meaning put on them that is wrong. Too often, 'participation' is reduced to public-awareness campaigns to promote acceptance of a prepared plan, or at best to canvass opinion about a range of options. 'Community involvement' is often downgraded to mean using free labour to build structures like canals, thus reducing construction costs. The attraction of 'farmer-managed irrigation', both in the context of indigenous irrigation and as a strategy for managing large-scale irrigation schemes, is often simply that it provides a way to reduce costs and avoid the corruption and inaction of government bureaucracies. Such approaches may be a little better than those that have gone before, but they are only tinkering with the edges of the problem.

Small-scale projects do not offer some clever new blueprint for development with which the experts of the aid/government machine can go on rolling into action. Rural Africans might need help, but not just any help. Community participation is not just another stage in project planning or a magical ceremony to guarantee success. If planning is centralized and based on the blueprints devised by outsiders, no form of words or hasty genuflection towards 'participation' (for example in the form of village meetings or 'public awareness campaigns') will ward off failure. The problems facing people in rural Africa cannot be solved that easily. The assumption that they can be only makes them worse. The 'solution' becomes part of the problem.

## Redesigning irrigation schemes

Interest in the development of indigenous irrigation in Africa is matched by a concern for the rehabilitation of existing large-scale irrigation schemes. Even though it is almost universally acknowledged that these have been failures, in most cases national governments see themselves as having little alternative to spending yet more money to try to make them work. Thus the Nigerian government is finishing the first 12,500 ha of the Hadejia Valley Project (after ten years in mothballs for lack of funds) using commercial loans from France, despite the probability of adverse downstream impacts and the likelihood that it will need to subsidized throughout its life.[14] The vast scale of the existing investment in infrastructure (in many cases built with loans that still have to be repaid) and the level of political rhetoric attached to large-scale irrigation make it difficult for governments to abandon them. Having attached emotive ideas about modernization and abundant future economic benefits to these projects, it is hardly feasible to turn around and deny them, although it can be done, as President Moi's scourging words about the Bura Project in Kenya in 1986 show. Even there, though, the project lingers as a dead weight around the neck of the Ministry of Agriculture and the national exchequer.

Rehabilitation is now the the fashionable subject among those interested in African irrigation. The FAO estimates that existing irrigation project rehabilitation might take place at a rate of about 50,000 ha per year.[15] Rehabilitation can mean re-designing the physical infrastructure of a project, for example to reduce problems of waterlogging and salinity in the ways described in Chapter 7. However, it can also involve re-organizing the administration and financial arrangements on a scheme, or changing the relations between scheme and farmers, or the ways farmers hold land. It may mean bringing in agribusiness management, or seeking new ways to involve farmers to a greater extent in scheme management. These things are very much more sensitive and complex. Nonetheless, there is now a clear consensus emerging about the need for reforms of this kind.

Unsurprisingly, perhaps, most of the new thinking about large-scale irrigation involves cutting bureaucracy and putting more management in the hands of farmers. A review of recent work on designs for sustainable farmer-managed irrigation in Africa highlighted three problems with government smallholder projects in Africa: 'achieving corporate identification and accountability on a non-kinship basis, managing money and managing equipment shared between more than one operator'.[16] In practice even the high cost of importing an outside management team is often ineffective, for example because of lack of local funds and constraints further down the bureaucracy, so a main conclusion was the need to delegate scheme management away from a centralized bureaucracy by vesting control in the hands of water users' committees. These cannot be created in a vacuum, and large investments of time and effort need to be spent in identifying existing community structure, power and leadership, if such associations are to have much chance of being effective.

This kind of approach has significant implications for design (or re-design) of irrigation systems. Instead of a single system that needs strong centralized management from the top of the main canal to the field level, a new 'system architecture' is required such that the scheme can be managed as a series of modules (not necessarily of the same size) that can operate semi-independently according to the decisions of the user group. Water application systems must be low-cost, and capital costs have to be low enough that when water is used on low-value local crops (or even fodder) an economic return remains possible.

New designs must take account of the need for easy local operation and maintenance, thus minimizing the risks associated with dependence on outside inputs of materials or expertize. Scheme design must also promote equitable distribution of water between modules and irrigators. Moreover this distribution must readily be seen to be equitable to avoid the technical inefficiencies in water use associated with cheating.

It seems to be widely assumed by development planners that equity is more easily achieved with farmer-control of development and management decisions. This may well be true, in that it reduces the scale of organization and makes decision-taking more accessible. However, it is easy to underestimate the persistence of inequalities in power and access to power. Probably the most important example of this is the position of women in irrigation. As Speelman comments, 'it is incorrect to assume that the farming family is a homogenous unit , with a single purse and with freely interchangeable or free family labour'.[17] Irrigation schemes in Africa that assume that farmers will be men, or that changes in cropping pattern, labour demand and farm economy will not affect the distribution of resources and labour between men and women, are unlikely to be successful (however clever they are in other ways). A study of a small Gambian irrigation scheme run by women points out that 'it is paradoxical that women, who are now recognized as major contributors to agricultural production in Africa, are victims of well-intentioned projects which undermine the very resource base on which their survival depends'.[18] Attempts to establish 'water user associations' without explicitly tackling the question of gender and guaranteeing voice and power to women are unlikely to be very successful.

New approaches to irrigation planning can take several forms. One study drawing on experience of village irrigation in the Senegal Valley calls for approaches to irrigation projects that 'reverse methods of design rather than vainly seek to adapt male and female smallholders to the designers' assumptions'.[19] What this seems to mean in practice is that feasibility assessment and design should include: first, a detailed description of production systems for different sectors (identifying whether male and female farmers see water as a key constraint); second, a detailed description of political culture and its implications for the design of infrastructure and organization of management; and third, detailed reports of negotiations between small farmers and officials as to production objectives, project layout and management. This agenda more or less stands normal professional practice on its head.[20] Few irrigation planners (almost all of whom are men) have got far in thinking these issues through in any formal way. Certainly there is a long way to go before such ideas become a standard part of training for irrigation engineers in either the North or the South.

Irrigation planning must also take account of local environmental and economic factors.[21] There must, for example, be specific provision for fuelwood supplies, and careful attention to the real economic conditions within which households work. Disruption of existing economic activities must be minimized. Particular attention must be paid to the integration of different economic activities (e.g. agriculture, livestock and fishing), the importance of off-farm incomes and the existence of problems such as seasonal shortages of labour. These issues can be complex and take time to unravel through painstaking research. If this is taken seriously it means that it will be extremely difficult to develop irrigation schemes rapidly.

Experience now shows clearly that in practice most people adopting irrigation fit it into a bundle of existing economic activities, rather than using it to replace them. Conflicts with those activities are therefore critical in determining whether irrigation is accepted or not, and whether it contributes to the magnitude or predictability of the returns which people obtain from floodplain resources. However high the potential yields, irrigation will not succeed unless it meets real needs. The reluctant adoption of the 'green revolution' swamp rice package in Sierra Leone (largely because of labour shortages which the package made worse not better) is a good example of this failure of an apparently clever package when introduced without proper understanding of socio-economic context.

In the past, irrigation schemes have generally been visualized as specialized islands of high technology, high yield (and high cost) production. Instead they need to be seen as part and parcel of the range of resources available, resilient to environmental changes, fitting flexibly into the changing economic demands of different kinds of households and responsive to the changing stresses and demands placed upon it. Irrigation needs to be understood not as a specialized activity which demands abandonment of other sources of income (e.g. herding or fishing), but as one among a portfolio of resources to be used to promote sustainable livelihoods. Irrigation development must cease to be treated as a single operation. Speelman argues that 'irrigation should be based on a concept that initiates a development process rather than (one) that plans a development action'.[22]

Irrigation schemes designed to guarantee high yields are characterized by detailed design, sophisticated technology, hierarchical and centralized planning. This approach to irrigation has given us projects that are too costly, too vulnerable to the natural and institutional environment of Africa, and too insensitive to the needs of water users. To be successful, irrigation must be very different: robust, flexible, even anarchic. Risk is endemic, and irrigation systems must be designed to work in an inherently risky environment, not to try to eliminate risk. Irrigation systems must be designed to run by themselves, and that means they must be designed by farmers. This is an obvious conclusion, and a drastically simple one. Nothing could be more radically different from the existing approaches to irrigation planning.

## Running the rivers

In just the same way that there is new thinking about irrigation in Africa, ideas about dams and floodplains are starting to change. It would be easy to conclude from the rather grim recital of the bad impacts of dam construction in Africa in this book that the whole business of river basin development is hopeless. The material is certainly available to do this quite well. You do not have to be a conspiracy theorist to see the stranglehold which the ideology of water development has on planning, or to unpack that ideology to see the vested interests and blinkered thinking within. This kind of argument would be easy to put forward. Had this book been written ten years ago I would probably have indulged myself. Perhaps I have grown less passionate in the interim, but for better or worse I have tried here to be pragmatic. And while a wholesale dismissal of dam construction is both possible and tempting, for a number of reasons it is not very sensible.

First, it fails to take account of the humane principle behind the urge to squeeze the natural environment within African countries in the hope of making life better for the poor. Decisions about dam construction in Africa are not being taken by fools, nor (in general) for perverse or corrupt reasons. Dams are intended to play a part in achieving developmental goals, most of which bear some relation to the kind of human welfare concerns which most people would share. The people of most African countries are facing conditions of poverty and exposure to hunger of a kind which demand solution, and no government worth anything can hide from that. It is not surprising that they turn to dams, and indeed to irrigation and every one of ninety-nine other supposed 'solutions' to their economic plight. They are paying good money for advice (and getting a great deal gratuitously from their bankers), and dams and river basin management form a central part of the wise words they receive. It is not particularly helpful to condemn the idea of river control and water resources planning simply because in practice so many plans conspicuously fail to work.

Second, a diatribe against dams is substantially a waste of breath. Their faults are clear enough, but however loudly proclaimed they are unlikely to prevail over the ideology of development which drives their creation. More rivers will be dammed and human control over Africa's water resources will increase. An Africa without dams is, like a world without nuclear weapons, probably beyond the reach of practical policy reform. It may be better to ride the tiger of developmentalism, and engage in debate to direct change, and to empower those affected by change to direct it for themselves, rather than stand back and try to stop the juggernaut of development in its tracks.

River valleys (like the rest of rural Africa) are subject to rapid change from outside pressures as they are brought into the modern world system. Economic integration with the national and international market, the cultural hegemony of the state (not least through education and the erosion of distance through improved transport) and the international spread of Westernized culture are creating enormous pressures on people and their resource use systems. Dams are one of these pressures, and while one of the more obviously drastic, are not isolated from the rest. Given the range and scope of the pressures for development, refusal to enter debates about how floodplain resources should be controlled and managed may prove to have high costs for the floodplain people currently most exposed to the negative aspects of change.[23]

Dams are something that Africa probably has to learn to live with. The question is how? One idea that has had a major impact on my thinking is provided by the work of Thayer Scudder on the potential for controlled flood releases from dams.[24] The idea is to use dams (both those already built and those still planned) for multiple purposes, one of which is to release an artificial flood to support agriculture, fishing and grazing in the downstream floodplain. Such a release would overcome the problem of unpredictablity in flood height, timing and duration which are undoubtedly major constraints on the productivity of floodplain land systems under uncontrolled conditions. This release of a known large amount of water at a predictable time would create a similarly predictable drawdown in the reservoir itself, which could be used for irrigation or grazing. The flood-release operation would therefore enable existing values of production in the downstream production to be maintained, and perhaps enhanced by appropriate low-input development to meet local constraints (e.g. credit, transport, extension advice and small-scale dykes and sluices), while the potential productivity of the reservoir for fish and other resources is maximized.

It is not easy to predict the exact volume and timing which would be optimal. The synchronization of water levels in different parts of the floodplain, and in the reservoir, would be complex, and the uncertainties of river flows in drought years would be as difficult as they are in 'conventional' river basin development. In order to make planned releases work, experimentation and good communications with people in the floodplain would be required. Moreover, the release might have to be of considerable magnitude in order to achieve adequate inundation in the downstream floodplain, and there would be costs associated both with the fact that the release would not be available for other purposes (e.g. HEP generation or formal irrigation outside the floodplain), and the need to reserve reservoir capacity to guarantee the release in low-flow years and protect the valley from excessive releases in high-flow years.

Thayer Scudder has discussed the controlled flood idea in a number of places now, including a USAID-funded study of African river basin planning and the 1988 UN/Economic Commission for Africa Interregional Meeting in Addis Ababa on river and lake basin development. The notion has been recommended by consultants for the management of the Jubba River following construction of the Bardeera Dam in Somalia. However, it has as yet rarely been tried out in practice.

One place where controlled flood releases have been tried is on the Pongolo River in Natal in South Africa.[25] The Pongolapoort Dam was constructed in the late 1960s for irrigation. It lies close to the Swaziland and Mozambique border, and the expected white settlers never materialized because of the fear of terrorists. The Natal Parks Board, University of Natal and Natal Provincial Fishery Institute got permission to experiment with controlled releases. The downstream floodplain is used by 25,000 Tembe-Thonga people for agriculture, grazing, fishing and gathering. However, flood management through the 1970s was planned by scientists primarily for fish resources, although the dependence of floodplain people on the flood was recognized, particularly following a seminar in 1978. In the meantime a small formal irrigation scheme of 2,700 ha was developed with Kwa-Zulu tenants. In 1989 a second workshop convened by the Department of Water Affairs discussed a revised release schedule based on studies of grazing and other human use. While the solution was a compromise, it was one in which local people were involved and their needs expressed and considered. Scudder comments, 'the process followed represented the type of suboptimization strategy which I believe is best suited to satisfy the needs of a multiplicity of users'.[26]

Of course, the Pongolo River case is unique in the sense that once built, the dam had no real purpose, and there is therefore no real competition for the water used for a downstream flood. Where this flood takes water from other uses, there will be economic trade-offs to be taken into account. Scudder argues that not only should the economics be worked out with great care, but that the poor performance of formal irrigation schemes and the uncertain added value following HEP generation might make the cost of the trade-off less than at first appears. If the full cost of losses of downstream production are taken into account, the balance of gains and losses may prove to swing in favour of the flood-release option.

An experiment carried out between 1988 and 1990 in the Senegal River may provide important evidence on this. The Manantali Dam on the Bafing River cost of the order of US $1 billion to build.[27] It has a significant impact on flooding in the lower Senegal, and effectively prevents recession cultivation on 100,000 ha of the Middle Valley. Although intended to allow formal irrigation of 250,000 ha on the Senegal bank of the river alone, neither Mauritania or Senegal had the capital to develop sufficient irrigated area following dam closure. The Organisation pour la Mise en Valeur du Fleuve Sénégal (OMVS) therefore built the dam with sufficiently large spillways to allow a flood-release. The Institute of Development Anthropology, based in the USA, has been studying the economics of existing production systems within the floodplain and the impacts of releases from the reservoir (and of course the uncontrolled flows of other tributaries of the Senegal).[28]

Experience with the release of an artificial flood has so far been mixed. The artifical flood of 7.5 $Mm^3$ was supposed to be released in September, to top up the flows in the unregulated Bakoye and Faleme rivers. In 1989 the operators at the dam were parsimonious, only releasing as much water from the dam as they calculated was arriving in the reservoir. The resulting flood flows were chaotic and small. In 1990 a decision was taken not to release the flood at all, although downstream users were not informed. The early warning system does not currently appear to be good enough to enable releases to be timed to enhance natural flood flows, even assuming those operating the dam were truly committed to the release, although there is no real reason why information flow should not be improved.

Initially the releases studied by IDA were intended to be short-term only, to allow time for the floodplain people to be 'weaned away from traditional agriculture to year-round irrigation'.[29] In practice, the rate of construction of modern irrigation systems has been much slower than predicted, and many of those that have been built have been abandoned by farmers within one or two years. IDA's work has suggested an alternative strategy, using an analysis of the costs and benefits of HEP generation and the artificial release. An engineering study done in 1987 argued that the artifical flood was incompatible with the demands of maximizing HEP generation. It calculated that the net benefits from flood-cropping in the valley were less than that from HEP generation, and therefore that water would be better used generating electricity than being released in an artifical flood. The IDA studies showed that some of the benefits of the flood had been underestimated, particularly fishing and livestock production. The inclusion of these tipped the balance of costs and benefits in favour of the artifical flood solution. It might be added to this that the economic inefficiency of the new irrigated villages perimeters (PIVs) supports the case for a more careful approach to floodplain development.[30]

Furthermore, IDA were able to model flows in the Senegal from 1904 to 1984 (a period including the severe droughts of the 1970s and 1980s), and calculate the effects of a flood release on HEP generation.[31] This study showed that 95 per cent of the time (925 out of 975 months) there would have been enough water to generate 74 MW of power and to release an artificial flood sufficient to inundate some 50,000 ha of land.[32] Only in the drought years of 1913, 1977 and 1979–84 was this impossible. IDA also examined the profound environmental impacts of curtailing the flood, including loss of floodplain woodland and a fall in the shallow water table used by villages within the floodplain, which is primarily recharged by floodwater.

It remains to be seen how the river basin planning mechanisms on the Senegal respond to the challenge of these radical ideas, both in terms of novel decisions about water use and in setting up effective mechanisms to communicate with users of released water hundreds of kilometres downstream. As Murdock and Horowitz point out, the decision is ultimately a political one. The costs to power generation of providing an artificial flood for downstream users are modest, while the costs of not doing so are considerable.

It will also be interesting to see how effectively the technicians and bureaucrats in the OMVS could design and monitor release patterns. Such tasks present novel and far from trivial professional challenges, and they are not ones with which donor agencies and the commercial companies that work for them have a great fund of experience. In the longer run, we might speculate as to whether such ideas might begin to undermine the ideology of river basin development which has held sway for so long, and what kind of backlash this might generate. Scudder himself is cautious about these issues. He points out that HEP benefits urban élites and industries whereas flood-releases benefit remote floodplain people. Not only that, but the economic benefits of the former are more easily measured (if only in electricity sales), and fit more easily into conventional models of industry-led development and rapid debt repayment.

A flood-release management strategy offers no easy way past the problems of limited hydrological knowledge which have dogged developments in the past. It requires a sophisticated hydro-meteorological data collection system capable of specifying in an accurate and timely manner the release flows needed. This task will be particularly difficult given the problems of drought and variable annual flows. There are also issues of disputes between upstream and downstream users, whose needs may conflict, and of conflicts between people using resources within the floodplain whose distribution and availability are changed by the new release patterns. Complex problems are also likely to emerge relating to the impacts of changed flooding regimes on the recharge of groundwater.[33] The flood-release strategy also, most obviously, demands dams with gates which allow large floods to be passed through the dam. Most existing dams lack such gates, and the potential for the approach in existing dams must be limited.

The use of controlled floods to convert single-purpose dams (e.g. for hydro-electric power generation) into a tool for multi-purpose multi-environment management, indeed the wider notion of using dams to work with the natural patterns in the rivers of Africa is attractive. Technically this idea is demanding since it requires both an understanding of downstream floodplain hydrology, effective real-time monitoring systems and an ability to make effective decisions and releases. However, there is no reason why African river basin agencies should not be capable (with training, support and suitable equipment) of tackling this. The transformation required in attitudes, in other words in ideology, is a more serious matter. River basin 'development' has to give way to river management. Development can no longer mean a single one-purpose project designed to work as an island in a sea of undeveloped bush. No longer can the 'success' of that project be assessed according to the performance within this area alone. Now development has to be seen as an ongoing process, and, moreover, one that demands involvement of different groups of people with different needs.

Such development not only has to be technically competent, it also has to be open and participatory, reflecting the needs and interests of different groups, not all of which will be compatible. Nor will the power and effective voice of different groups necessarily be related to their need for floodplain resources. These attributes of controlled flood management are not going to fit easily into established patterns of development planning, which are dominated by centralization and narrowly technical issues, and by the urban élites who comprise the planners.

Ian McHarg wrote a book in 1969 called *Design with Nature*.[34] It advocated landscape design in tune with natural processes and patterns. It is a principle which has attracted increasing support in Europe and North America through the environmental revolution of the 1970s and 'greening' of the 1980s. It has rarely been tried in African rivers and wetlands. There, the old watchwords of large-scale development and environmental transformation still hold the field. Perhaps their failures will finally bear them down, but how are they to be replaced?

## Managing Africa's wetlands

There are alternatives to large-scale transformation of Africa's rivers and floodplain wetlands. One is the challenge from a perhaps surprising source, the international conservation movement. The International Union for the Conservation of Nature and Natural Resources (IUCN, or World Conservation Union) has a Wetlands Programme which is now initiating a series of innovative projects in partnership with African governments. Their approach builds on the principles of sustainable development in the World Conservation Strategy published by IUCN, the World Wildlife Fund (WWF; now known as the Worldwide Fund for Nature) and the United Nations Environment Programme in 1980,[35] and the principles of 'wise use' of wetlands adopted under the Ramsar Convention on wetlands.[36] The central principle is the ecological productivity of wetlands and the attempt to identify a role for them at the centre of sustainable development strategies. As Patrick Dugan points out, this is not the approach used in the past: 'Rather than devise development strategies which focus on the productivity of these ecosystems and respond to the needs of the rural producers who depend upon them, most development investment has centred on agricultural and urban expansion to meet the needs of urban populations'.[37] One result of this has been that rural people have been deprived of the resources on which they depend.

The IUCN Wetlands Programme is working in a series of tropical African countries. In Uganda, they have helped the government develop a National Wetlands Conservation and Management Programme (through the Ministry of Environment Protection), and work is being done to promote sustainable fishery development in the Doho Swamps in eastern Uganda and to determine the limits of sustainable harvesting of Papyrus in Nakyeteme Swamp West of Kampala. In Tanzania, with help from IUCN, the National Environment Management Council has begun to work with other government agencies, NGOs and academic institutions to design a national wetland conservation and management programme. In Nigeria, IUCN is involved as a partner in the Hadejia-Nguru Wetlands Conservation Project, an attempt to promote integrated and sustainable use of the extensive floodplains of the Hadejia and Jama'are rivers against pressures of upstream water abstraction, drought and demands for canalization downstream.

Similar initiatives have been undertaken by other agencies: for example, in Zambia a project designed by IUCN and funded and implemented by the Worldwide Fund for Nature in collaboration with the Department of National Parks and Wildlife has been running since 1986 to promote the conservation and management of the natural resources of the Kafue Flats and Bangewelu Swamps. In Mali IUCN is assisting the Ministries of Environment and Livestock to investigate indigenous resource use in the Niger Inland Delta, and then to develop micro-projects to try to demonstrate sustainable development options. In Cameroon IUCN is assisting the Ministry of Planning and other concerned government ministries to examine the feasibility of rehabilitating the floodplain of the Logone River, and integrating such a rehabilitation programme within the regional land use planning process in Cameroon.

Some conservationists see African floodplains as needing protection from people and conservation management to promote the interests of wildlife. The IUCN approach is getting away from this, at least on paper, to focus on the economic uses of floodplains and the ecological and physical processes that sustain them. An attempt is being made to relate the working of wetland ecosystems to the resources they can provide, and to focus in particular on the rights and needs of floodplain people. This is an important emphasis.

Of course, in some areas wildlife may prove to be a valuable element in the future economics of wetlands, particularly if there is thought to be potential for wildlife tourism. Thus the strategy in the Kafue Flats includes a scheme to harvest the Kafue Lechwe (a small wetland antelope) such that revenues from controlled hunting will feed back to local people. Experience with strategies such as this, gained where there is potential conflict between pastoralists and wildlife such as the Maasai at Amboseli in Kenya, is mixed. Financial flows are generally small, slow in coming and the links between preservation and investment can be unclear. However, such participation in wildlife utilization has proved effective in Zimbabwe, and on the Kafue Flats in Zambia has already led to a reduction in unlicenced hunting and an increased commitment to maintain floodplain resources among what was a depressed and deprived community. In some wetlands it is likely that wildlife conservation will have a role to play in resource development and bringing improved incomes for the rural poor.

Elsewhere, other approaches will be needed to address the needs of those depending on floodplain resources. The important thing about the IUCN approach is its focus on the ways natural systems work, the way people have adapted to them, and the strategies necessary to enhance the productivity of the resources in ways that maximize human benefit. There is also a new focus on *how* development is to be planned. The new ideas involve considerable shifts in attitude for environmentalists moving from Northern conservation to Southern people-orientated environmental concern. They also offer a challenge to the ways that development planners have long tackled their task. Perhaps with the clear sight born of naïvety, environmentalists are defining a far more humane approach to development: 'wetland "management" consists of providing people with the capacity to solve their own problems, and the administrative and land-tenure structures which will allow them to use these, rather than some higher authority (either government or expatriate projects) trying to do so for them'.[38]

This new agenda for 'conservation' action has much to teach the development community. It is not particularly new, since ideas about participation in development, 'bottom-up' planning and (more recently) empowerment are all established parts of the development planning debate. However, it is timely, and the ideas are being adopted with an enthusiasm that is extremely refreshing in the battle-weary world of African development. It is the implementation of these vague wish-words that is the challenge needing to be faced, and here the attempt to implement these ideas under the banner of 'wetland conservation' is challenging and instructive.

The importance of the experiment is being recognized by the willingness of donor agencies to commit funds to these new projects. Finnish and Norwegian aid, for example, is funding work on wetland management in the SADCC countries of southern Africa, and part of the funding for the WWF/Zambia Wetlands Project comes from Britain's Overseas Development Administration. Dutch aid is supporting work on floodplain rehabilitation in Cameroon and the Uganda national wetlands programme. That programme was previously funded by NORAD. Of course, to an extent, Northern aid agencies are simply being driven by public opinion at home to fund 'environmentally sustainable' projects and thus to reflect the 'greening' of government, but this does not mean that their interest is not timely and important. The interests of wildlife conservationists in the North can have even more direct impacts on project development, for example in the WWF 'debt-swap' arrangement with Zambia that is helping provide funds for the Zambia Wetlands Project. Here, as in other cases, the broad umbrella of 'sustainable development' is breaking down the once-clear divisions between 'conservation' and development. The outcome of this curious evolution in thinking is still unclear, but it certainly provides an approach to the people and environments of African floodplains that is very different from the engineering-centred approach of the past.

# Network planning: experts helping experts

There is obviously a need to re-design the way we approach development. There are fundamental flaws both in the way development projects are planned, and also, more fundamentally, in the way development itself is conceived. The very notion of science and technology, of development planning and expenditure, as ways to remove 'obstacles' to economic growth, needs to be looked at with some scepticism. The priorities set by such a view of Africa need to be compared with the probability of achieving them, and with simple human priorities. The risks of established styles of development need to be compared with the possible benefits. There must be a greater scepticism about the capacity of Northern science to second-guess nature in Africa, and caution must replace gung-ho confidence in the ability of technology and engineering to remove constraints on rural production and on the livelihoods of rural people.

There must obviously be far more respect for indigenous knowledge, and recognition of the importance of diversity and flexibility as responses to the fickle environment of Africa. Most indigenous resource management systems have these characteristics: most development projects do not. What Camilla Toulmin and Robert Chambers call 'third' agriculture (by which they mean not the industrialized agriculture of the North or the productive 'green revolution' agriculture of the South) has fared very badly under the development initiatives of the last half century.[39] It is complex, diverse and risk-prone, but it is also resilient and remarkably productive. As E.F. Schumacher wrote on the eve of his death in 1977, the Third World poor are survival artists: 'it is quite certain that if there should be a real resource crisis, or a real ecological crisis, in this world, these people will survive. Whether you or I will survive is much more doubtful. India will survive, though whether Bombay will survive is more doubtful. That New York will survive is an impossibility'.[40]

There must also be far greater understanding of the social and political implications of development interventions, of gender and community politics. This requires a great deal more research, and (more importantly) a great deal of learning and self-education by those who presume to come to Africa (or come to rural Africa from the cities) to bring 'development'. In many instances the best place to start this research/learning process will be by listening to farmers, women as well as men, and giving them the chance to speak.

There is a need for humility and for experiment in the face of uncertainty, lack of facts and lack of resources – and, it should be added, uncertain bureaucratic competence. The notion that development planners can offer a blueprint for development has to change. Development must be seen as a learning process,[41] 'a process of experimentation which should permit the project in hand to evolve at a pace suited to human and environmental circumstances, establishing by trial and error new patterns of economic and social behaviour'.[42]

Robert Chambers identifies a series of problems in development that stem from the way outsiders normally work and learn. Normal professionalism, the dominant way of doing things, 'tends to put things before people, men before women, the rich before the poor and the urban and industrial before the rural and agricultural'.[43] Normal bureaucracy is hierarchical and tends to centralize, standardize and regulate. Normal careers related to rural life 'often start at the periphery and then move upwards in hierarchies and inwards to larger and larger urban centres'.[44] Normal learning is from 'above', from teachers, books and urban centres of knowledge, from capital cities and from the First World. It is not from the rural poor.

In response to these problems, Chambers suggests a strategy of reversals, of 'turning the normal on its head'. In terms of professionalism, it means putting the last first, 'permitting and promoting the complexity and diversity the poor people want, presenting them with a basket of choices rather than a package of practices'.[45] Bureaucratically, it means destandardizing, decentralizing power and removing restrictions. In careers it means rewarding those who listen and those who search for new ideas and solutions.[46]

One element in the reversals necessary to change the way development planning is done involves the people doing it. Chambers comments that 'the neglect of personal attitudes and behaviour has been a stunning oversight in rural development practice'.[47] Development planning runs on the views of outside 'experts', people paid to identify 'problems', and come up with plausible solutions. People doing this work are often insulated from the subject of their analysis and the impacts of their recommendations by lifestyle, just as they are separated from the people on whose behalf they work by experience. Their expertize is therefore of a certain very limited kind. They have training in science, economics and engineering. They have money with which to pay people to collect data, and they have the training to analyse it. They may well also have insights and understanding that can be important and much needed. Unfortunately, they also tend to have prejudices and cultural preconceptions that are no help at all. As a result, their prescriptions are constrained within the narrow boundaries of the known.

This works in simple disciplinary terms. Engineers come up with engineering solutions. Agronomists come up with solutions involving methods of crop husbandry familiar in the North, based on inorganic fertilizer and agrochemicals. It also works in a wider sense. Experts from the First World make assumptions about social structure and gender, about what people want from life and what 'development' ought to mean. Unconsciously, they recreate the wage-earning lifestyle of Northern suburbia in the Africa bush. Experts are also selectively blind. They do not see the grip of their own economies and cultures. They do not see the risks, in human, economic and environmental terms, associated with the remedies they prescribe. Above all, these experts are too often willing to think narrowly, at least in their professional (if not their personal) lives. They concentrate on a single goal (for example, increasing agricultural production) and try to devise ways to meet it, assuming that in this way development will be fostered.

But these outsiders are not the only experts around. The farmers, fishermen and herders for whom they labour are also experts. They also have experience and training. They also have knowledge of the way the environment changes, of the workings of the economy in which they are situated, of the technical constraints on different kinds of production. Of course, this knowledge is not encyclopaedic, but then nor is that of the outside expert. The knowledge of the local expert is, indeed, deficient in just those areas in which the outsider's is best. They may not have a full understanding of long-term environmental or social trends, or be able to predict future changes in national or international economy. Yet they certainly know the importance of these things in a way that the outsider never will. Furthermore, they have no difficulty in thinking broadly. Life is not divided into 'problems' to be solved logically in isolation from each other. Subsistence and family, economy and community, income and health: all these things are inseparable. 'Problems' do not exist in isolation, they are part of life. And 'development' is not some abstract goal, it is the reality of tomorrow and the day after, a future for the generations to come.

Development planning, therefore, cannot be done through clever blueprints. It cannot be done outside rural communities, however diligently the fruits of that planning are aired and discussed with rural people afterwards. People need to be allowed to make their own decisions about their future. If they do not, then the money spent on development will probably be wasted, and its effects will be at best irrelevant to peoples' lives and at worst disastrous. But people cannot work alone. Indigenous resource use systems in Africa are under pressure from the world economy, from globalized Western culture, from population growth and from climatic change. People in Africa need tools and resources to swim against these pressures. To speak of 'traditional' methods as if they could survive unchanged in the world at the end of the twentieth century is as cruel as it is historically false. Societies, economies and cultures will continue to change: the question is how, and who will control that change. Rural Africans need support, and they certainly need resources. They need technology, and they need some relief from corrupt and warring governments and the agendas of international business. What they do not need is simplistic recipes for change that offer them no choice and no control.

Fortunately, the outside experts who run the development business also have knowledge of a different kind. The engineer or agriculturalist who goes to an African village and sits down and talks brings with him or her much more than a knowledge of their own discipline. First, they bring their humanity. They are people, just as are those to whom they speak. Talking face to face as people has enormous potential. It is a deeply subversive act, opening a vast kaleidoscope of possibilities. The outside expert steps outside the armour of his or her expertize, and in so doing discovers the possibility of listening as well as talking, of learning as well as teaching and becomes that most human of beings, a visiting stranger. Human contact and human trust are the only basis for action that will endure. Secondly, the visitor brings knowledge of the wider world. This brings risks (for example, the cultural pollution of a rich Northerner or a city-dwelling African) but also vast possibilities. The visitor is a gatekeeper, knowing what resources are available and how to access them. The visitor is a prophet, able to see beyond the horizons of the village and go some way at least towards explaining that larger world. The visitor is an opening into a larger network of other people with other knowledge and other expertize: all of them people, all of them capable of humanity if they will come and listen and talk.

## Making it happen

Development planning is not an arena for expert action, for dramatic actions and sweeping solutions. It is not a field where experts meet peasants and prescribe treatments, like some medical team beamed down to a strange planet from the Starship Enterprise. If development is to mean anything, it must mean change that people can understand and can control. It means the end of 'blueprint planning', whether it be a grandiose plan to control water flow in a river basin or a small-scale plan to redesign channels on an indigenous irrigation scheme. Instead, it demands what I would call 'network planning', ideas and innovations flowing not down from above but up and across.

In network planning local and outside experts would work together, sharing knowledge and generating action, reflecting upon it and responding to it at each moment. Change would be steered continuously, not produced in a series of drastic thrusts. There must be continuity over time: outsiders cannot jet in and jet out and expect to effect lasting and positive change. Policy makers must grow towards the communities they work with, becoming part of them just as they are part of the network for development that they help create. In so doing, they must transform the institutions within which they work, from dictatorial hierarchical agencies offering top-down development planning into agencies that facilitate and coordinate the work of those they are established to serve.

It should not be thought that such 'development' will be easy. Career structures and systems of bureaucratic organization and reward are geared to hierarchy, loyalty and convention. Network planning, and the reversals Robert Chambers calls for, challenge this. The human relations on which network planning is based are costly and long-lasting. There are also no simple criteria for judging success, and few rewards can be expected from the government ministries or donor agency bureaucracies within which network planners work. There are professional risks for individuals, just as there are in any other field where a new way of thinking, a new paradigm, starts to challenge the old. Network planning is humane, in a world that seeks to make decisions that are logical and meet narrowly economic goals. It works by empowering the powerless, which is the reverse of the way the world generally works.

Network planning demands changes of attitude, but it also demands institutional reform. First, in order to give adequate attention to the need to learn from and with the rural poor of Africa's floodplains, development agencies and institutions will have to redefine and restructure corporate and professional goals and ideologies. Issues of project and programme coordination must be addressed, and new approaches to wetland environments and resources must be put in place to match new relations with floodplain people.

Second, development institutions and national agencies need to accept that change will be slower, more gradual and less predictable than they would probably like. Patrick Dugan argues that this must be 'accepted as a necessary condition of effective interaction with rural communities'.[48] To meet this challenge, donors will have to accept longer-term commitments to particular areas and particular institutions.

They will have to accept that the most effective investments are not always the easiest to label and claim credit for.

Third, agencies and governments must realize that effective use of rivers and wetlands will require careful appraisal of wetland environments and functions and their capacity to support intensive management of different kinds.[49] Full environmental appraisal must become an integral part of development planning, in practice as well as in theory, and must be fostered by the public policy environment. This must also tackle issues of rights and security of tenure for the rural poor, and must support conciliation and conflict resolution. This policy environment must also enable institutions to tackle the information, intervention and market failures that threaten the sustainability of wetland livelihoods.[50] Taxes, subsidies and quotas that promote large-scale intervention in wetland environments must be replaced with others that foster the interests of the poor.

Institutional reform is necessary to allow new approaches to the people and environments of Africa. It will, by a nice irony, only come about through the adoption of those appoaches and the personal attitudes that drive them. However, neither new attitudes nor institutional reform are enough. Rural Africans are not free agents, to be released so easily from the constraints that bind them. The reform of institutions and attitudes must be matched by real empowerment and by changes in the structures within which rural communities are locked. This change will be brought about not by well-meaning strangers, but by what Anthony Bebbington calls 'everyday organized local struggle'.[51] The ability of the poor to secure rights to property and rights to floodplain resources will be a key factor in determining the future shape of floodplain development. Development agencies, development institutions and individuals can and should seek to support the poor in their struggles for rights and resources.

None of this makes a very neat agenda for change, nor is it an easy one. The development of African rivers and wetlands is not something that can be achieved through some single all-out effort, nor is it something that can be done through some Grand Plan involving the dramatic application of 'Northern' knowledge or methods. The future lies within rural Africa itself, in the hands of farmers and pastoralists. They need the chance of obtaining the power to build that future for themselves and the opportunity to choose its shape. There is urgent work to be done in empowering and sustaining their efforts. There is an important role for those with the eyes to see and the humility to listen, learn and share.

# Further reading

## Chapter one: Introduction

Grove, A.T. (1989) *The Changing Geography of Africa*, Oxford University Press, Oxford.

Maltby, E. (1988) *Waterlogged Wealth: Why waste the world's wet places?* Earthscan, London.

## Chapter two: Changing Africa

Dumont, R. (1962) *Afrique Noire est Mal Partie*, Editions de Seuil, Paris. (Published in English 1966 as *False Start in Africa* by André Deutsch, London, and reprinted 1988 by Earthscan, London.)

Grove, A.T. (1989) *The Changing Geography of Africa*, Oxford University Press, Oxford.

Harrison, P. (1987) *The Greening of Africa: breaking through in the battle for land and food*, Paladin, London.

Lewis, L.A. and Berry, L. (1988) *African Environments and Resources*, Unwin Hyman, London.

O'Connor, A. (1991) *Poverty in Africa: a geographical approach*, Belhaven, London.

Rimmer, D. (ed.) (1991) *Africa 30 Years On*, Royal African Society, James Currey, London and Heinemann, Portsmouth.

Timberlake, L. (1985) *Africa in Crisis: the causes, the cures, of environmental bankruptcy*, Earthscan, London.

World Bank (1990) *Sub-Saharan Africa, from Crisis to Sustainable Growth: a long-term perspective study*, World Bank, Washington.

## Chapter three: The hard earth

Barrow, C. (1988) *Water Resources Development in the Tropics*, Longman, London.

Chambers, R. (1983) *Rural Development: putting the last first*, Longman, London.

Grainger, A. (1982) *Desertification: how people make deserts, how people can stop and why they don't*, Earthscan, London.

Grove, A.T. (ed.) (1985) *The Niger and its Neighbours: environmental history and hydrobiology, human use and health hazards of the major West African rivers*, Balkema, Rotterdam.

Grove, A.T. (1989) *The Changing Geography of Africa*, Oxford University Press, Oxford.

Roberts, N. (1988) *The Holocene*, Basil Blackwell, Oxford.

Warren, A. and Agnew C. (1988) *An Assessment of Desertification and Land Degradation in Arid and Semi-Arid Areas*, IIED Dryland Programme Paper No. 2, November 1988.

# Chapter four: Using Africa's wetlands

Drijver, C.A. and M. Marchand (1985) *Taming the Floods: environmental aspects of floodplain development in Africa*, Centre for Environmental Studies, State University of Leiden.

Dugan, P.J. (ed.) (1990) *Wetland Conservation: a review of current issues and required action*, International Union for the Conservation of Nature, Gland, Switzerland.

Grove, A.T. (ed.) *The Niger and its Neighbours: environment history and hydrobiology, human and health hazards of the major West African rivers*, Balkema, Rotterdam.

Richards, P. (1985) *Indigenous Agricultural Revolution*, Hutchinson, London.

Richards, P. (1986) *Coping with Hunger: hazard and experiment in a West African rice-farming system*, Allen and Unwin, Hemel Hempstead.

Scoones, I. (1991) 'Wetland in Drylands: key resources for agricultural and pastoral production in Africa', *Ambio* 20:366–371.

Sutton, J.E.G. (ed.) (1989) 'History of African Agricultural Technology and Field Systems', *Azania* 24:1–122.

Welcomme, R.L. (1979) *Fisheries Ecology of Floodplain Rivers*, Longman, London.

# Chapter five: Dreams and schemes: planning river development

Adams, W.M. (1985) 'River basin planning in Nigeria', *Applied Geography* 5:297–308.

Collins, R.O. (1990) *The Waters of the Nile: hydropolitics and the Jonglei Canal 1900–1988*, Clarendon Press, Oxford.

Godana, B.A. (1985) *Africa's Shared Water Resources: legal and institutional aspects of the Nile, Niger and Senegal River Systems,* Lynne Rienner, Boulder, Co.

Saha, S.K. and Barrow, C.J. (1981) (eds) *River Basin Planning: theory and practice,* Wiley, Chichester.

Salau, A.T. (1986) 'River basin planning as a strategy for rural development in Nigeria', *J. Rural Studies* 2(4):321–335.

Scudder, T. (1989) The African experience with river basin development', *Natural Resources Forum* May 1989:139–148.

Scudder, T. (1989) 'River Basin Projects in Africa: conservation vs. development', *Environment* 31(2)4–32.

Scudder, T. (in press) *The African Experience with River Basin Development,* Clark University and Institute for Development Anthropology (to be published by Westview Press).

Waterbury, J. (1979) *Hydropolitics of the Nile Valley,* Syracuse University Press, Syracuse, NY.

## Chapter six: Binding the rivers

Adams, W.M. (1990) *Green development: environment and sustainability in the Third World,* Routledge, London.

Collins, R.O. (1990) *The Waters of the Nile: hydropolitics and the Jonglei Canal 1900–1988,* Clarendon Press, London.

Grove, A.T. (ed.) *The Niger and its Neighbours: environment, history, hydrobiology, human use and health hazards of the major West African Rivers,* Balkema, Rotterdam.

Hansen, A. and Oliver-Smith, S.A. (1982) *Involuntary migration and resettlement: the problems and responses of dislocated people,* Westview Press, Boulder, Co.

Howell, P.P., M. Lock and S. Cobb (eds) *The Jonglei Canal: impact and opportunity,* Cambridge University Press, Cambridge.

Payne, A.J. (1990) *The Ecology of Tropical Lakes and Rivers,* Wiley, Southampton.

Welcomme, R.L. (1970) *Fisheries Ecology of Floodplain Rivers,* Longman, London.

## Chapter seven: Watering the savanna

Adams, W.M. and A.T. Grove (eds) *Irrigation in Africa: problems and problem-solving*, Cambridge African Monograph No. 3 African Studies Centre, Cambridge.

Barrow, C.J. (1989) *Water Resource Development in the Tropics*, Longman, London.

Chambers, R. (1989) *Managing Canal Irrigation*, Cambridge University Press, Cambridge.

Moris, J.R. and Thom, D.J. (1990) *Irrigation Development in Africa: lessons of experience*, Westview Press, Boulder Co.

Underhill, H. (1990) *Small Scale Irrigation*, Cranfield Press, Silsoe.

# Notes

## Chapter one: Introduction

1. Zechariah 4 verse 10, Authorised Version, Holy Bible.
2. World Bank (1990) *Sub-Saharan Africa: from Crisis to Sustainable Growth: a long-term perspective study*, World Bank, Washington, p. 27.
3. Goulet, D. (1971) *The Cruel Choice: a new concept in the theory of development*, Atheneum, New York.
4. Maltby, E., (1988) *Waterlogged Wealth: Why waste the world's wet places?* Earthscan, London p. 28.
5. See for example Maltby, E., *op. cit.;* and Williams, M. (1990) *Wetlands; a threatened landscape*, Basil Blackwell, Oxford.
6. Turner, K. (1991) 'Economics and wetland management', *Ambio* 20: 59–63.
7. Dugan, P. J. (ed.) (1990) *Wetland Conservation: a review of current issues and required actions*, International Union for the Conservation of Nature, Gland, Switzerland.
8. Welcomme, R.L. (1979) *Fisheries Ecology of Floodplain Rivers*, Longman, London.
9. The agroecology of wadis in Sudan is described by M. O. El Sammani (1991) 'Wadis of North Kordofan: present roles and prospects for development', in I. Scoones (ed.) *Wetlands in Drylands: the agroecology of savanna systems in Africa* Part 3c, IIED Drylands Programme, London.
10. See for example I. Scoones (ed.) (1991) *Wetlands in Drylands: the agroecology of savanna systems in Africa, Part 1*, IIED Drylands Programme, London.
11. Moorehead, R. (1988) 'Access to resources in the Niger Inland Delta, Mali, pp. 27–39 in J.A. Seeley and W.M. Adams (eds) *Environmental Issues in African Development Planning*, Cambridge African Monographs 9, African Studies Centre, Cambridge.
12. The story of the Volta River Project is described in Hart, D. (1980) *The Volta River Project: a case study in politics and technology*, Edinburgh University Press, Edinburgh.
13. Grove, A.T. (1991) 'The African environment', pp. 39–55 in D. Rimmer (ed.) *Africa 30 Years On*, Royal African Society, James Currey, London and Heinemann, Portsmouth.

## Chapter two: Changing Africa

1. Barra, G. (1960) *Kikuyu Proverbs, with translations and English equiv-alents*, 2nd. Ed., Kenya Literature Bureau, Nairobi.
2. The 'myth of primative Africa', see Hopkins, A.G. (1973) *An Eco-nomic History of West Africa*, Longman, London. p. 10.
3. See for example the discussion and the call for a comparative social history in Iliffe, J. (1987) *The African Poor: a history*, Cam-bridge University Press, Cambridge.
4. Rodney, W. (1972) *How Europe Underdeveloped Africa*, Bogle-L'Ouverture Publications, London.
5. Kjekshus, H. (1977) *Ecology Control and Economic Development in East African History*, University of California Press, Berkeley. Another author developing a similar theme (for Zambia) is Vail, L. (1977) 'Ecology and history: the example of Eastern Zambia', *Journal of South African Studies* 3:129–155.
6. Ford, J. (1971) *The Role of Trypanosomiasis in African Ecology: a study of the tsetse fly problem*, Clarendon Press, Oxford. The interpretation of Ford's book is reviewed in J. Giblin (1990) 'Trypanosomiasis control in African history: an evaded issue', *Journal of African His-tory* 31:59–80, and the complexities of the interactions between tsetse fly, ecology and history are explored in Waller, J. (1990) 'Tsetse fly in Western Narok, Kenya', *Journal of African History* 31:81–101.
7. Hopkins, A.G. (1973) *An Economic History of West Africa*, Longman, London, p. 10.
8. The importance of ideas of tropical 'Edens' in relation to Euro-pean exploration is developed by R.H. Grove (1987) 'Early themes in African conservation: the Cape in the Nineteenth Century', pp. 21–40 in Anderson, D.M. and Grove, R.H. (eds) *Conservation in Africa: people, policies and practice*, Cambridge University Press, and (1990) 'The origins of environmentalism', *Nature* 345 (6270): 11–14. Also interesting is A. Graham (1973) *The Gardeners of Eden*, Allen and Unwin, Hemel Hempstead.
9. Hopkins *op. cit.* p. 10.
10. McCracken, J. (1987) 'Colonialism, capitalism and the ecological crisis in Malawi: a reassessment', pp. 63–77 in Anderson, D.M. and Grove, R.H. (eds) *Conservation in Africa: people, policies and practice*, Cambridge University Press.
11 Iliffe, J. (1978) 'Farm and fly in Tanganyika', *Journal of African History* 19:139–141. See also Iliffe, J. (1981) *A Modern History of Tanganyika*, Cambridge University Press.

12. Watts, M.J. (1983) *Silent Violence: food, famine and the peasantry in northern Nigeria*, University of California Press, Berkeley; see also Watts, M.J. 'The demise of the moral economy: food and famine in the Sudano-Sahelian region in historical perspective', pp. 128–148 in E.P. Scott (ed.) *Life Before the Drought*, Allen and Unwin, Hemel Hempstead.

13. Franke, R.W. and Chasin, B.H. (1980) *Seeds of Famine: ecological destruction and the development dilemma in the West African Sahel*, Allenheld and Osman, Montclair, NJ.

14. Crosby, A. (1986) *Ecological Imperialism: the biological axpansion of Europe 900–1900*, Cambridge University Press, Cambridge.

15. Gorer, G. (1935) *Africa Dances*, republished Penguin Books 1945, Harmondsworth; reprinted 1983.

16. Weiskel, T. (1988) 'Towards an archaeology of colonialism: elements in the ecological transformation of the Ivory Coast', pp. 141–171 in D. Worster (ed.) *The Ends of the Earth: perspectives on modern environmental history*, Cambridge University Press, Cambridge.

17. Wrigley, C.C. (1989) 'Bananas in Buganda', *Azania* 24:64–70.

18. A good introduction to this new environmental history is D. Worster (ed.) (1988) *The Ends of the Earth: perspectives on modern environmental history*, Cambridge University Press, Cambridge. The approach is developed in the African context by D. Johnson and D.M. Anderson (eds) (1988) *The Ecology of Survival: case studies from Northeast African History*, Lester Crook, London and Westview Press, Boulder, Co.

19. Dumont, R. (1962) *L'Afrique Noire est Mal Partie*, Editions de Seuil, Paris (published in English 1966 as *False Start in Africa* by André Deutsch, London, and reprinted 1988 by Earthscan, London. Quotes are pp. 32–33).

20. There is an excellent (although depressing) review of African development in A. O'Connor's book *Poverty in Africa: a geographer's perspective* (1991), Belhaven, London.

21. These and other figures in this chapter are drawn from World Bank (1990) *Sub-Saharan Africa: from Crisis to Sustainable Growth: a long-term perspective study*, World Bank, Washington.

22. Research by SADCC, ECA and UNICEF, quoted in World Bank (1990) *op. cit.*, p. 23.

23. Cervenka, Z. (1989) 'The relationship between armed conflicts and environmental degradation in Africa', pp. 25–36 in A. Hjort af Örnas and M.A. Mohammed Salih (eds) *Ecology and Politics: environmental stress and security in Africa*, Scandinavian Institute of African Studies, Uppsala.

24. O'Connor, A. (1991) *Poverty in Africa: a geographer's perspective*, Belhaven, London. p. 55.
25. World Bank (1981) *Accelerated Development in Sub-Saharan Africa*, World Bank, Washington.
26. Hjort af Örnas, A. (1989) 'Environmental land security of dryland herders in East Africa', pp. 67–88 in A. Hjort af Örnas and M.A. Mohammed Salih (eds) *Ecology and Politics: environmental stress and security in Africa*, Scandinavian Institute of African Studies, Uppsala.
27. This account draws extensively on M. Lipton and R. Longhurst (1989) *New Seeds and Poor People*, John Hopkins University Press, Baltimore.
28. Richards, P. (1985) *Indigenous Agricultural Revolution: ecology and food production in West Africa*, Hutchinson, London. The changes needed in plant breeding research are discussed in A. Haugerud and M.P. Collinson (1991) *Plants, Genes and People: improving the relevance of plant breeding*, IIED Gatekeeper Series 30, IIED Sustainable Agriculture Programme, London.
29. Chambers, R. (1983) *Rural Development: putting the last first*, Longman, London.
30. For example J.C. Stone (ed.) (1980) *Experts in Africa*, African Studies Centre, University of Aberdeen.
31. I should say that as an academic in a First World University I am acutely aware that such a double meaning applies very much to my own case!
32. Adams, A. (1979) 'An open letter to a young researcher', *African Affairs*, 78:451–479.
33. Among many general accounts, see M. Howes and R. Chambers (1979) 'Indigenous technical knowledge: analysis, implications and issues', *IDS Bulletin* 10(2):5–11; and D.M. Warren (1991) *Using Indigenous Technical Knowledge in Agricultural Development*, World Bank Discussion Paper 127, Washington.
34. This memorable phrase comes from P. Richards (1985) *Indigenous Agricultural revolution: ecology and food production in West Africa*, Longman, London.
35. I am grateful to Tony Bebbington for these insights.
36. Sutton, J.E.G. (1990) *A Thousand Years of East Africa* British Institute in Eastern Africa, Nairobi.
37. Chambers, R. (1991) 'In search of professionalism, bureaucracy and sustainable livelihoods for the 21st Century', *IDS Bulletin* 22(4):5–11.

# Chapter three: The hard earth

1. de Schlippe, P. (1956) *Shifting Agriculture in Africa: the Zande System of Agriculture*, Routledge and Kegan Paul, London, p. 12.
2. A compendium of maps of rainfall variability in Africa is provided by Nicholas, S.E., Kim, J. and Hoopingarner, J. (1988) *Atlas of African Rainfall and its Variability*, Department of Meteorology, Florida State University, Tallahassee, Fa.
3. There is a good general review in Niewolt, S. (1982) *Tropical climatology*, Wiley, Chichester. There are accounts of the particular problems on West Africa in Hare, F.K. (1983) 'Climate on the desert fringe', pp. 134–151 in R. Gardner and H. Scoging (eds) *Mega-geomorphology*, Clarendon Press, Oxford, and in Hayward, D.F. and Oguntoyinbo, J.S. (1986) *Climatology of West Africa*, Hutchinson, London.
4. Farmer, G. (1986) 'Rainfall variability in tropical Africa: some implications for policy', *Land Use Policy* 3:336–342.
5. Trilsbach A. and Hulme, M. (1987) 'Recent rainfall changes in central Sudan and their physical and human implications', Transactions of the Institute of British Geographers N.S. 9: 280–298.
6. Hulme, M. (1987) 'Secular changes in wet season structure in central Sudan, *Journal of Arid Environments* 13:31–46.
7. Kimmage, K. (1991) 'Small-scale irrigation initiatives in Nigeria: the problems of equity and sustainability', *Applied Geography* 11: 5–20.
8. Lipton, M and Longhurst, R. (1989) *New Seeds for Poor People*, J. Hopkins University Press, Baltimore. The issue of plant breeding is debated in Chapter 2.
9. The use of *guna* is described by M. Mortimore (1989) *Adapting to Drought: farmers, famine and desertification in West Africa*, Cambridge University Press, Cambridge.
10. Sinclair, A.R.E. and Norton-Griffiths, M. (eds) *Serengeti: dynamics of an ecosystem*, University of Chicago Press, Chicago.
11. Breman, H. and de Wit, C. (1983) 'Rangeland productivity and exploitation in the Sahel', *Science* 221: 1341–1347.
12. Western, D. (1982) 'The environment and ecology of pastoralists in arid savannas', *Development and Change* 13:183–211.
13. Stenning, D.J. (1959) *Savanna Nomads, A study of the Wodaabe pastoral Fulani at Western Borno Province, Northern Region, Nigeria*, International African Institute and Oxford University Press, London, p. 207.
14. Robinson, P.W. (1989) 'Reconstructing gabbra history and chronology: time reckoning, the Gabbra calendar and the cyclical view

of life', pp. 151–168 in T.E. Downing, K.W. Gitu and C.M. Kamau (eds) *Coping with Drought in Kenya; national and local strategies*, Lynne Rienner, Boulder, Co. and London, p. 161.

15. Swift, J. (1982) 'The future of African hunter-gatherer and pastoral peoples', *Development and Change* 13:159–181.

16. Dyson-Hudson, N. and Dyson-Hudson, R. (1982) 'The structure of East African herds and the future of East African herders', *Development and Change* 13:233–238.

17. This work is described in M.B. Coughenour *et al.* (1985) 'Energy extraction and use in a nomadic pastoral ecosystem', *Science* 230(4726)619–625, and also in Coppock, D.L. *et al.* (1986) 'Livestock feeding ecology and resource utilisation in a nomadic pastoral ecosystem', *Journal of Applied Ecology* 23:573–589.

18. Pankhurst, R. and Johnson, D.H. (1988) 'The great drought and famine of 1882–92 in northeast Africa', pp. 47–72 in Johnson, D. and Anderson, D. (eds) *The Ecology of Survival: case studies from northeast African history*, Lester Crook, London and Westview Press, Boulder, Co.

19. Waller, R. (1988) 'Emutai: crisis and response in Maasailand 1883–1902', pp. 73–112 in Johnson, D. and Anderson, D. (eds) *op. cit.*. See also Waller, R. (1990) 'Tsetse fly in Western Narok, Kenya', *Journal of African History* 31:81–101.

20. Waller, R. (1990) 'Tsetse fly', p. 100.

21. Hjort af Örnas, A. (1989) 'Environment and security of dryland herders in east Africa', pp. 67–88 in A. Hjort af Örnas and M.A. Mohammed Salih (eds) *Ecology and Politics: environmental stress and security in Africa*, Scandinavian Institute of African Studies, Uppsala.

22. This view is quoted, and the general issue of attitudes to pastoralists is discussed in R.C. Bridges (1991) 'Official perceptions during the colonial period of problems facing pastoral societies in Kenya', pp. 141–153 in J.C. Stone (ed.) *Pastoral Economies in Africa and Long-Term Responses to Drought*, Aberdeen University African Studies Group, Aberdeen.

23. Horowitz, M. and Little, P.D. (1987) 'African pastoralism and poverty: some implications for drought and famine', pp. 59–82 in M.H. Glantz (ed.) *Drought and Hunger in Africa: denying famine a future*, Cambridge University Press; see also Swift (1982), *op. cit.*

24. Bridges *op. cit.*

25. Quoted in Mace, R. (1991) 'Overgrazing overstated', *Science* 399 (24 January 1991):280–1.

26. Timberlake, L. (1985) *Africa in Crisis: the causes, the cures of environmental bankruptcy*, Earthscan, London.

27. See, for example, Grove, A.T. (1991) 'The African environment', pp. 39–55 in D. Rimmer (ed.) *Africa 30 Years On*, Royal African Society, James Currey, London and Heinemann, Portsmouth; and Lewis, L.A. and Berry, L. (1988) *African Environments and Resources*, Unwin Hyman, London. There are many accounts of individual environmental problems, notably perhaps fuelwood, for example G. Leach and R. Mearns (1988) *Beyond the Woodfuel Crisis: people, land and trees in Africa*, Earthscan, London; and Munslow, B., Katerere, Y., Ferf, A. and O'Keefe, P. (1988) *The Fuelwood Trap: a study of the SADCC Region*, Earthscan, London.

28. Grove, R.H. (1987) 'Early themes in African conservation: the Cape in the nineteenth century', pp. 21–39 in Anderson, D.M. and Grove, R.H. (eds) *Conservation in Africa: people, policies and practice*, Cambridge University Press, Cambridge.

29. For example Stebbing, E.P. (1938) 'The encroaching Sahara: the threat to the West African colonies', *Geographical Journal* 85: 506–24.

30. Jones, B. (1938) Desiccation in the West African Colonies, *Geographical Journal* 91:401–423.

30. Furon, R. (1947) *L'Erosion du Sol*, Payot, Paris.

32. Anderson, D.M. (1984) 'Depression, dust bowl, demography and drought: the colonial state and soil conservation in East Africa in the 1930s, *African Affairs* 83:321–344.

33. Adams, W.M. (1987) 'Approaches to water resource development in the Sokoto Valley, Nigeria: the problem of sustainability', pp. 307–325 in Anderson, D.M. and Grove R.H. (eds) *op. cit.*

34. Aubréville, A. (1959) *Climâts, Forêts et Desertification de l'Afrique Tropicale*, Société d'Editions Géographiques, Maritimes et Coloniales, Paris.

35. See for example Grove, A.T. (1977) 'Desertification', *Progress in Physical Geography* 1:296–310, and also United Nations (ed.) (1977) *Desertification: its causes and consequences*, Pergamon Press, Oxford.

36. Warren, A. and Agnew, C. (1988) *An Assessment of Dryland Desertification and Land Degradation in Arid and Semi-Arid Areas*, Drylands Programme Paper no. 2, International Insitute for Environment and Development, London.

37. The problems of defining desertification are discussed in detail by M.M. Verstrate (1986) 'Defining desertification: a review', *Climatic Change* 9:5–18, and also by M. Mortimore (1989) *op. cit.*

38. Walls, J. (1984) 'Summons to action', *Desertification Control Bulletin* 10:5–14.

39. Reviews of what is known about desertification include L. Berry (1984) 'Desertification in the Sudano-Sahelian region 1977–1984',

*Desertification Control Bulletin* 10:23–28 and M.K. Tolba (1986) 'Desertification in Africa', *Land Use Policy* 3: 260–268.

40. Tucker, C.J., Townshend, J.R.G. and Goff, T.E. (1985) 'African land-cover classification using satellite data', *Science* 227(4683): 369–375; Dregne, H.E. and Tucker, C.J. (1988) 'Green biomass and rainfall in semi-arid sub-Saharan Africa', *Journal of Arid Environments* 15:245–252.

41. Tucker, C.J., Justice, C.O. and Prince, S.D. (1986) 'Monitoring the grasslands of the Sahel 1984–1985', *International Journal of Remote Sensing* 7:1571–1583.

42. Helldén, U. (1988) 'Desertification monitoring: is the desert encroaching?', *Desertification Control Bulletin* 17:8–12.

43. See Warren, A. and Agnew, C. (1988) *op. cit.*

44. Quoted in Behnke, R. and Scoones, I. (1991) *Rethinking Range Ecology: implications for rangeland management in Africa*, Overview of Paper presentations and Discussions at the technical meeting on Savanna Development and Pasture production 19–21 November 1990, Woburn, UK, Commonwealth Secretariat, London.

45. IUCN Sahel Programme (1989) *Sahel Information Kit*, IIED Drylands Network Paper 14, London.

46. Warren, A. and Agnew, C. (1988) *An Assessment of Desertification and Land degradation in Arid and Semi-Arid Areas*, IIED Drylands Network Paper No. 2, London.

47. Adams, M.E., personal communication.

48. Lamb, P.J. (1982) 'Persistence of sub-Saharan drought', *Nature* 299:46–47.

49. Farmer, G. (1989) *Rainfall in the Sahel*, IUCN Sahel Programme, IIED Drylands Network Paper 10, London.

50. Grove, A.T. (1973) 'A note of the remarkably low rainfall of the Sudan Zone in 1913', *Savanna* 2:133–8.

51. Agnew, C. (1990) 'Spatial aspects of drought in the Sahel', *Journal of Arid Environments* 18:279–293.

52. Nicholson, S.E. (1978) 'Climatic variations in the Sahel and other African regions during the last five centuries', *Journal of Arid Environments* 1:3–24.

53. Grove, A.T. and Warren, A. (1968) 'Quaternary landforms and climate on the South side of the Sahara', *Geographical Journal* 134: 194–208; and Nichol, J. (1991) 'The extent of desert dunes in northern Nigeria as shown by image enhancement', *Geographical Journal* 157:13–24.

54. Thomas, D.S.G. (1984) 'Ancient ergs of the former arid zones of Zimbabwe, Zambia and Angola', *Transactions of the Institute of British Geographers* N.S. 9:75–89. The Quaternary history of the

Kalahari is reviewed in Thomas, D.S.G., and Shaw, P. (1991) *The Kalahari Environment*, Cambridge University Press.

55. Grove, A.T. and Warren, A. (1968) *op. cit.*

56. Grove, A.T. (1985) 'The physical evolution of the river basins', pp. 21–60 in A.T. Grove (ed.) *The Niger and its neighbours: environmental history and hydrobiology, human use and health hazards of the major West African rivers*, Balkema, Rotterdam.

57. Moore, P.D. (1990) 'Palaeoecology: ups and downs in the Sahel', *Nature* 343 (6527):414.

58. Haynes, C.V., Eyles, C.H., Pavlish, L.A., Ritchie, J.C. and Rybak, M. (1989) 'Holocene palaeoecology of the eastern Sahara: Selima Oasis', *Quaternary Science Reviews* 8: 109–136.

59. Street, F.A. and Grove, A.T. (1976) 'Environmental and climatic implications of late Quaternary lake-level fluctuations in Africa', *Nature* 261: 385–90.

60. Lézine, A-M. and Casanova, J. (1989) 'Pollen and hydrological evidence for the interpretation of past climates in tropical West Africa during the Holocene', *Quaternary Science Reviews* 8:45–55.

61. Grove, A.T. (1985) *op. cit.*

62. Lézine and Casanova (1989) *op. cit.*

63. Tyson, P.D. (1986) *Climatic Change and Variability in Southern Africa*, Oxford University Press, Cape Town.

64. This is particularly associated with the work of J. Otterman (1974) 'Baring high albedo soils by overgrazing: a hypothesised desertification mechanism', *Science* 166:531–533; and J. Carney (1975) 'Dynamics of deserts and drought in the Sahara', *Quarterly Journal of the Royal Meteorological Society* 101:193–202.

65. This work is reviewed by J.G. Lockwood (1983) 'The influence of vegetation on the earth's climate', *Progress in Physical Geography* 7: 81–96 and S.E. Nicholson (1988) 'Land surface atmosphere interaction: physical processes and surface changes and their impact', *Progress in Physical Geography* 12:36–55.

66. Folland, C.K., Palmer, T.N. and Parker, D.E. (1986) 'Sahel rainfall and worldwide sea surface temperatures, 1901–1985', *Nature* 320: 602–607.

67. Street-Perrott, F.A. and Perrot, R.A. (1990) 'Abrupt climatic fluctuations in the tropics: the influence of Atlantic Ocean circulation', *Nature* 343:607–612.

68. Dennett, M.D., Elson, J. and Rodgers, J.A. (1985) 'A reappraisal of rainfall trends in the Sahel', *Journal of Climatology* 5:353–361.

69. Sutcliffe, J.V. and Lazenby, J.B.C. (1990) 'Hydrological data requirements for planning Nile management', pp. 107–136 in P.P. Howell and Allen J.A. (eds) *The Nile: resource evaluation, resource*

*management, hydropolitics and legal issues,* School of Oriental and African Studies and Royal Geographical Society, London.

70. Grove, A.T. (1991) 'The African environment', pp. 39–55 in D. Rimmer (ed.) *Africa: 30 years On,* Royal African Society, James Currey, Heinemann, London.

71. Charnock, A. (1985) 'Lake Nasser empties as drought spreads North', *New Scientist,* 2 May 1985, p. 7.

72. Farmer (1986) 'Rainfall variability in tropical Africa: some implications for policy', *Land Use Policy* 3:336–342, p. 342.

73. Evans, T. (1990) 'History of Nile flows', pp. 5–39 in P.P. Howell and J.A. Allen (eds) *op. cit.*

74. Sutcliffe and Lazenby (1990) *op. cit.*

75. Kolawole, A. (1987), Environmental change and the South Chad Irrigation Project (Nigeria), *Journal of Arid Environment* 13:169–176.

# Chapter 4: Using Africa's wetlands

1. Richards, P. (1988) 'The versatility of the poor: indigenous wetland management systems in Sierra Leone', Paper to the African Studies Association Conference, Cambridge, September 1988 (in press, G. Elwert, ed.).

2. This phrase is taken from the title of a research project carried out by the International Institute for Environment and Development (IIED) Drylands Programme 'Wetlands in Drylands; Agroecology of Savanna Ecosystems in Africa'. This is published in three parts (8 volumes), see I. Scoones (ed.) (1991) *Wetlands in Drylands: Agroecology of Savanna Ecosystems in Africa, Part 1: Overview – ecological, economic and social issues,* IIED Drylands Programme, London; and Scoones, I. (1991) 'Wetlands in Drylands: key resources for agricultural and pastoral production in Africa', *Ambio* 20:36–371.

3. Richards, P. (1985) *Indigenous Agricultural Revolution: ecology and food production in West Africa,* Hutchinson, London; and (1986) *Coping with Hunger: hazard and experiment in a West African rice-farming system,* Allen and Unwin, Hemel Hempstead.

4. Léricollais, A. and Schmitz, J. (1984) 'La Calabasse et la houe: techniques et outils des cultures de décrue dans la vallée du Sénégal', *Cahiers ORSTOM Sér. Sci. Hum.* 20:427–452.

5. Richards, P. (1983) Farming systems and agrarian change in West Africa, *Progress in Human Geography,* 7(1):1–39.

6. Marzouk, Y. (1989) 'Sociétés rurales et techniques hydrauliques en Afrique', *Études Rurales* 115/6:9–36.

7. Rydzewski, J.R. (1987) 'Introduction: Planning of irrigation development' pp. 1–10 in J.R. Rydzewski (ed.) *Irrigation Development Planning*, Wiley, Chichester.

8. This point is developed in W.M. Adams and D.M. Anderson (1988) 'Irrigation before development: indigenous and induced change in agricultural water management in East Africa', *African Affairs* 87:519–535.

9. Sutton, J. (1989) 'Towards a history of cultivating the fields', *Azania* 24: 98–112.

10. Marzouk, Y. (1989) *op. cit.*

11. Carter, R. (ed.) (1988) Newsletter Number 1 of the Informal Working Group on Small-Scale Irrigation, June 1988, p. 3.

12. FAO Investment Centre (1986) *Irrigation in Africa South of the Sahara*, Food and Agriculture Organisation Investment Centre Technical Paper 5, Rome.

13. Johnny, M., Karimu, J. and Richards, P. (1981) 'Upland and swamp rice farming in Sierra Leone: the social context of technical change', *Africa*, 51:596–620 p. 603.

14 Seignobos, C. (1984) 'Instruments aratoires du Tchad Méridional et du Nord-Cameroun', *Cahiers ORSTOM Sér. Sci. Hum.* 20:537–575.

15. Richards, P. (1988) 'The versatility of the poor: indigenous wetland management systems in Sierra Leone', Paper to the African Studies Association Conference, Cambridge, September 1988 (in press, G. Elwert, ed.)

16. Richards, P. (1989) 'Agriculture as performance', pp. 39–43 in R. Chambers, A. Pacey and L.A. Thrupp (eds) *Farmer First: farmer innovation and agricultural research*, I.T. Publications, London.

17. Richards, P. (1986) *Coping with Hunger: hazard and experiment in a West African rice-farming system*, Allen and Unwin, Hemel Hempstead.

18. See for example, R. Portères (1970) 'Primary cradles of agriculture on the African continent', in J. Fage and R. Oliver (eds) *Papers in African Prehistory*, Cambridge University Press.

19. Macintosh, S.K. and Mackintosh, R.J. (1980) *Prehistoric investigations in the region of Jenne, Mali*, Cambridge Monographs in African Archaeology 2, British Archaeological reports, Oxford; See also McIntosh, R.J. and McIntosh, S.K. (1984) 'Early iron age economy in the Inland Delta (Mali)', pp. 158–172 in J.D. Clark and S.A. Brandt (eds) *From Hunters to Farmers*, University of California Press, Berkeley Ca. The more general archaeological background is re-

viewed in G. Connah (1987) *African Civilisations: precolonial cities and states in tropical Africa, an archaeological perspective*, Cambridge University Press, Cambridge.

20. Connah, G. (1981) *Three Thousand Years in Africa: man and his environment in the Lake Chad region of Nigeria*, Cambridge University Press, Cambridge.

21. Maiga, A., de Leuw, B.N., Diarra, L., and Hiernaux, P. (1991) 'The harvesting of wild-growing grain crops in the Gourma Region of Mali', IIED Drylands Network Issues Paper No. 27.

22. Marzouk-Schmitz, Y. (1984) 'Instruments aratoires, systèmes de cultures et différentation intra-ethnique', *Cah. ORSTOM Sér. Sci. Hum.* 20 (3-4):399-425.

23. R. Jobson (1623) *The Golden Trade. or A Discovery of the river Gambia, and the golden trade of the Aethiopians*, N. Okes, London (reprinted 1968), Dawsons of Pall Mall, London. I am grateful to Paul Richards for this and other references in this chapter.

24. This account draws on A. Léricollais and J. Schmitz (1984) "'La Calabasse et la Houe": techniques et outils des cultures de décrue dans la vallée du Sénégal', *Cahiers ORSTOM Séri. Sci. Hum.* 20 (3-4):427-452; J-L Boutillier and J. Schmitz (1987) 'Gestion traditionelle des terres (système de Décrue/système pluvial) et transition vers l'irrigation' *Cahiers Sciences Humaine* 23:533-554; J. Schmitz (1986) 'Agriculture de décrue, unités térritoriales et irrigation dans la vallée du Sénégal', *Les Cahiers de Recherches Development* 12:65-77.

25. Indigenous floodplain agriculture in the Sokoto Valley is described in Adams W.M. (1986) 'Traditional agriculture and water use in the Sokoto Valley, Nigeria', Geographical Journal 152:30-43.

26. Seignobos, C. (1984) 'Instruments aratoires du Tchad Méridional et du Nord-Cameroun', *Cah. ORSTOM Ser. Sci. Hum.* 20 (3-4): 537-573.

27. Grove, A.T. (1985) 'Water characteristics of the Chari system and Lake Chad', in A.T. Grove (ed.) *The Niger and its Neighbours: environmental history and hydrobiology, human use and health hazards of the major West African rivers*, Balkema, Rotterdam pp. 21-60.

28. Connah, G. (1985) 'Agricultural intensification and sedentism in the Firki of northeast Nigeria', in Farrington I.S. (ed.) *Prehistoric Intensive Agriculture in the Tropics*, BAR International series 232, Volume II.

29. Kolawole, A. (1991) 'Economics and management of fadama in northern Nigeria', in I. Scoones (ed.) *Wetlands in Drylands: the agroecology of savanna systems in Africa*, IIED, London.

30. Richards, P. (1988) 'The versatility of the poor: indigenous wetland management systems in Sierra Leone', Paper to the African Studies Association Conference, Cambridge, September 1988.

31. Adams, W.M. (1990) 'Definition and development in Africa indigenous irrigation', *Azania* 24: 21–27.

32. Scudder, T. (1962) *The Ecology of the Gwembe Tonga,* Manchester University Press, Manchester, for the Rhodes-Livingstone Institute, Kariba Studies Volume II.

33. Soper, R.C. (ed.) *Socio-cultural profile of Turkana District,* Institute of African Studies, Nairobi, and Ministry of Finance and Planning, Government of Kenya.

34. Dickie, A. (1991) 'Systems of agricultural production in the Southern Sudan', pp.280–307 in G.M. Craig (ed.) *The Agriculture of the Sudan,* Oxford, University Press; See also Howell, P., Lock, M. and Cobb, S. (eds) *The Jonglei Canal: impact and opportunity,* Cambridge University Press, Cambridge.

35. Floodplain cultivation in the Inland Niger Delta is described in several places, notably by Jean Gallais, for example Gallais J. (1967) *La Delta Intérieure du Niger: étude de geographie régionale,* IFAN, Dakar; and (1984) *Hommes du Sahel: La Delta Intérieure du Niger 1960–1980,* Flammarion, Paris. There are accounts in English by Harlan, J.R. and J. Pasquerau (1969) 'Décrue agriculture in Mali', *Economic Botany* 23: 70–74 and Scudder, T. (1980) 'River basin development and local initiative in Savanna environments', in D.R. Harris (ed.) *Human Ecology in Savanna Environments,* Academic Press, London.

36. Gallais, J, and Sidikou, H.H. (1978) 'Traditional strategies, modern decision-making and management of natural resources in the Sudan-Sahel, pp. 11–34 in MAB (ed.) *Management of Natural resources in Africa: traditional strategies and modern decision-making,* MAB Technical Notes 9, UNESCO, Paris.

37. Littlefield, D.C. (1981) *Rice and Slaves: ethnicity and the slave trade in colonial South Carolina,* Louisiana University Press, Baton Rouge and London.

38. Linares, O.F. (1981) 'From tidal swamp to inland valley: on the social organisation of wet rice cultivation among the Diola of Senegal, *Africa* 51:557–595; also Y. Marzouk-Schmitz (1984) 'Instruments aratoires, systèmes de cultures et differentation intra-ethnique', *Cahiers, ORSTOM Sér, Sci. Hum.* 20 (3-4): 399–425.

39. Chisholm, N.G. and Grove, J.M. (1985) 'The lower Volta', pp. 229–250 in A.T. Grove (ed.) *The Niger and its Neighbours: environmental history and hydrobiology, human use and health hazards of the major West African rivers,* Balkema, Rotterdam.

40. Turner, B. (1986) The importance of dambos in African agriculture, *Land Use Policy* 3:343–347; Bell, M. and Roberts, N. (1991) 'The political ecology of dambo soil and water resources in Malawi', *Transactions of the Institute of British Geographers N.S.* 16:301–318.

41. See for example C. Raynaut (1989) 'La culture irriguée en pays Haussa Nigerién: aspects historiques, sociaux et techniques', *Études Rurales* 115/6:105–128.

42. Turner, B. (1984) 'Changing land use patterns in the fadamas of northern Nigeria' pp.140–170 in E.P. Scott (ed.) *Life Before the Drought*, Allen and Unwin, Hemel Hempstead.

43. Connah, G. (1987) *African Civilisations: precolonial cities and states in tropical Africa, an archaeological perspective*, Cambridge University Press, Cambridge.

44. de Sainte Marie, C. (1989) 'État et paysans dans les systèmes hydrauliques de la vallée du Nil (Égypte)', *Études Rurales* 115/6: 59–91.

45. Dafalla, H. (185) *The Nubian Exodus*, C. Hurst and Co., London; also R.T. Wilson (1991) 'Systems of agricultural production in northern Sudan', pp. 280–307 in G.M. Craig (ed.) *The Agriculture of the Sudan*, Oxford University Press.

46. Bernus, E. (1989) 'L'eau du désert: usages, techniques et maîtrise de l'espace aux confins du Sahara', *Études Rurales* 115/6: 93–104.

47. Gallais, J, and Sidikou, H.H. (1978) 'Traditional strategies, modern decision-making and management of natural resources in the Sudan-Sahel', pp. 11–34 in MAB (ed.) *Management of Natural resources in Africa: traditonal strategies and modern decision-making*, MAB Technical Notes 9, UNESCO, Paris.

48. Bernus, E. (1989) *op. cit.*

49. Meschy, L. (1989) 'La colline et le marais: la gestion des bassins versants au Burundi et au Rwanda', *Études Rurales* 115/6: 129–151.

50. Henning, R.O. (1941) 'The furrow-makers of Kenya' *Geographical Magazine* 12: 268–279; E. Huxley (1959) 'African water engineers', *Geographical Magazine* 32: 170–175.

51. Soper, R. (1983) 'A survey of the irrigation systems of the Marakwet', pp. 96–111 in Kipkorir B.E., R. Soper and J.W. Ssennyonga (eds) *Kerio Valley: past, present and future*, Nairobi.

52. Sutton, J.E.G. (1984) 'Irrigation and soil conservation in African agricultural history: with a reconsideration of the Inyanga terracing (Zimbabwe) and Engaruka irrigation works (Tanzania)', *Journal of African History* 25:25–41.

53. Hallpike, C.R. (1970) 'Konso agriculture', *Ethiopian Studies* 8 (1): 31–44; H. Amborn (1989) 'Agricultural intensification in the Burji-Konso cluster of southwestern Ethiopia', *Azania* 24: 71–83.

54. Lowe-McConnell, R.H. (1985) 'The biology of the river systems with particular reference to the fishes', pp. 101–140 in A.T. Grove (ed.) *The Niger and its Neighbours: environmental histroy and hydrobiology, human use and health hazards of the major West African rivers,* Balkema, Rotterdam.

55. Welcomme, R.L. (1979) *Fisheries Ecology of Floodplain Rivers,* Longman, London. There has been extensive research on the Kafue Flats by the Kafue Basin Research Project of the University of Zambia, for example Williams, G.J. and Howard, G.W. (eds) (1977) *Development and Ecology in the Lower Kafue Basin in the Nineteen Seventies,* Kafue Basin Research Committee, Lusaka, Zambia; and Handlos, W.L. and Williams, G.J. (1984) *Development on the Kafue Flats: the last Five Years,* Kafue Basin Research Committee, Lusaka, Zambia.

56 Welcomme, R.L. (1979), *op. cit.*

57. Matthes, H. (1990) Report on the Fishery-related Aspects of the Hadejia-Nguru Wetlands Conservation Project, Mission Report, HNWCP, Nguru, Nigeria; see also Kimmage, K. and Adams, W.M. *The Value of Economic Production in the Hadejia-Jama'are Floodplain, Nigeria,* Report to IUCN Wetlands Programme.

58. Moorehead, R. 'Access to resources in the Niger Inland Delta, Mali', pp. 27–39 in J. Seeley and W.M. Adams (eds) *Environmental Issues in African Development Planning,* African Studies Centre Cambridge, Cambridge African Monographs 9; and Moorehead, R. (1990) 'Land tenure and environmental conflict: the case of the Inland Niger Delta, Mali', unpublished paper, March 1990.

59. Stiles, D. (1990) *Lakes of Grass: regenerating bourgou in the Inner Delta of the Niger River,* United Nations Sudan-Sahelian Office (UNSO) Technical Paper 2, New York.

60. Drijver, C.A. and Marchand, M. (1985) *Taming the Floods: environmental aspects of floodplain development in Africa,* Centre for Environmental Studies, University of Leiden.

61. This account is based on R. Moorehead (1990) 'Land tenure and environmental conflict: the case of the Inland Niger Delta, Mali', unpublished paper, March 1990.

# Chapter five: Dreams and schemes: planning river development

1. Stamp, L.D. (1953) *Our Undeveloped World*, Faber and Faber, London.
2. Bates, M. (1953) *Where Winter Never Comes: a study of man and nature in the tropics*, Victor Gollancz, London, p. 118.
3. Bates, M. *op. cit.*, p. 239.
4. D.A. Low and J.M. Lonsdale (1976) 'Towards a new order 1945–1963', pp. 1–63 in D.A. Low and J.M. Lonsdale (eds) *History of East Africa* Volume III, Clarendon Press, Oxford, p. 13.
5. D.A. Low and J.M. Lonsdale (1976), *op. cit.*, p. 14.
6. Worthington, E.B. (1983) *The Ecological Century: a personal appraisal*, Cambridge University Press, Cambridge.
7. Culwick, A.T. (1943) 'New beginning', *Tanganyika Notes and Records* 15: 1–6, p.5.
8. Stamp, L.D. (1938) 'Land utilisation and soil erosion in Nigeria', *Geographical Review* 28: 32–45.
9. Gourou, P. (1948) *Les Pays Tropicaux*, Presses Universitaires de France, St. Germain, Paris; pp. 111–113 ('une civilisation plus evoluée riche d'éléments empruntés a la Mediterranée').
10. Roberts, P. (1981) '"Rural Development" and the rural economy in Niger, 1900–1975', pp. 193–221 in J. Heyer, P. Roberts and G. Williams (eds) *Rural Development in Tropical Africa*, Macmillan, London.
11. Pélissier, P. and Diarra, S. (1978) *Traditonal strategies, modern decision-making and management of natural resources in Sudan Africa*, UNESCO MAB Technical Note 9 (Management of Natural Resources in Africa: traditional strategies and modern decision-making) UNESCO, Paris.
12. Jones, G.H. (1936) '*The Earth Goddess: a study of native farming on the West African Coast*', Longman, Green and Co., London.
13. Baldwin, K.D.S. (1957) *The Niger Agricultural Project: an experiment in African development*, Basil Blackwell, Oxford, p. 1.
14. Allen, F. (1983) 'Natural resources as national fantasies', *Geoforum* 14: 243–247
15. Commonwealth Development Corporation (1952) *Report on the Gambia Egg Scheme*, Cmd. 8560, May 1952.
16. Fiah. E. (1982) 'Why the groundnut scheme failed', *Tanzania Daily News*, 16 August 1982.
17. Baldwin *op.cit.*, p. 9.
18. Baldwin *op.cit.*, p. 10.
19. All quotes from Baldwin *op. cit.* pp. 11–12.

20. Fyfe, C. (1980) 'Agricultural dreams in Nineteenth Century West Africa', p.11 in J.C. Stone (ed.) *Experts in Africa*, Aberdeen University.

21. Rydzewski, J.R. (1997) 'Introduction: planning of irrigation development', pp. 1–10 in J.R. Rydzewski (ed.) *Irrigation Development Planning*, Wiley, Chichester.

22. Anderson, D.M. (1988) 'Cultivating pastoralists: ecology and economy among the Il Chamus of Baringo 1840–1980', pp. 241–260 in D. Johnson and D.M. Anderson (eds) (1988) *Ecology and Survival: case studies from northeast African history*, Lester Crook, London; see also D.M. Anderson 'Agriculture and irrigation technology at Lake Baringo in the Nineteenth Century', *Azania* 24: 84–97.

23. Chambers, R. (1973) 'The Perkerra Irrigation Scheme: a contrasting case', pp. 344–364 in Chambers R. and Moris, J. (eds) *Mwea: An irrigated rice settlement in Kenya*, Weltforum Verlag, Munich; p.347. There is also an account of the origins of the Perkerra scheme in Anderson 'Cultivating Pastoralists', *op.cit.*

24. Anderson 'Cultivating Pastoralists', *op. cit.*, p. 258.

25. Obara, D.A. (1984) 'Irrigation schemes in arid environments of Kenya, with special reference to Perkerra Irrigation Scheme', pp. 201–207 in M.J. Blackie (ed.) *African Regional Symposium on Small Holder Irrigation*, Hydraulics Research Ltd, Wallingford. The performance of this and other Kenya schemes is reviewed in W.M. Adams (1990) 'How beautiful is small? scale, control and success in Kenyan irrigation', *World Development* 18(10): 1309–1323.

26. Richards, P. (1986) *Coping with Hunger: hazard and experiment in an African rice-farming system*, Allen and Unwin, London.

27. Richards, P. *ibid.*, p. 7.

28. Richards, P. *ibid.*, p. 8

29. Carney, J. and M. Watts (1990) 'Manufacturing dissent: work, gender and the politics of meaning in a peasant society', *Africa* 60:207–230.

30. Richards, P. (1986) *op. cit.*, p. 22.

31. The development of the Gezira scheme is described in A. Gaitskell (1959) *Gezira: a story of development in the Sudan*, Faber and Faber, London; a good critique of its many problems is T. Barnett (1977) *The Gezira Scheme: an illusion of development*, Frank Cass, London. The production system is described in I.G. and M.C. Simpson (1991) 'Systems of agricultural production in central Sudan and Khartoum Province', pp. 252–279 in G.M. Craig (ed.) *The Agriculture of the Sudan*, Oxford University Press.

32. Babiker, A.B.A.G. (1982) 'The Gezira Scheme: development problems of a modern large-scale irrigated experiment in arid Africa,'

pp. 85–95 in H.G. Mensching (ed.) *Problems of the Management of Irrigated Land in Areas of Traditonal and Modern Cultivation*, IGU Working Group on Resource Management in Drylands, Hamburg.

33. Baldwin *op. cit.*, p. 12.

34. de Wilde, J.C. (1967) *Experiences with Agricultural Development in Tropical Africa: Volume II, the case studies*, Johns Hopkins Press, Baltimore, for the International Bank for Reconstruction and Development.

35. Harrison Church, R.J. (1961) 'Problems of the development of the dry zone of West Africa', *Geographical Journal* 127: 187–204.

36. Bortoli, L. and G. Sournia (not dated) 'Les mirages de l'irrigation et le sous-developpement: cas de l'Afrique sèche de l'Quest,' (unpublished manuscript).

37. World Water (1989) 'Optimising water supplies on a grand scale', *World Water* July/August 1989:35–6.

38. Pearce, F. (1991) 'Africa at a watershed', *New Scientist*, 23 March 1991: 34–39.

39. Umolu, J.C. and Oke, V.O. (1986) 'Zaire-Chad-Niger Interbasin Water Transfer Scheme: a proposal for subregional water resources planning', paper to International Conference on *Water Resources Needs and Planning in Drought-prone Areas*, Khartoum 6–12 December 1986.

40. Umolu and Oke *op. cit.*, p. 3.

41. The scheme is discussed in W.M. Adams and G.E. Hollis (1989) *Hydrology and Sustainable Resource Development of a Sahelian Floodplain Wetland*, Hadejia-Nguru Wetlands Conservation Project, London.

42. Bonifica SpA (1988) *Transaqua: a North-South Idea for South-South Cooperation*, pp. 5–41.

43. Scudder, T., (1989) 'River basin projects in Africa: conservation vs. development, *Environment* 31(2): 4–32, p. 8

44. The history of Nile management is described in Waterbury, J. (1979) *Hydropolitics of the Nile Valley*, University of Syracuse Press, Syracuse NY, and R.O. Collins (1990) *The Waters of the Nile: hydropolitics and the Nile 1900–1988*, Clarendon Press, Oxford.

45. For an account of the planning of the Jonglei canal, see Waterbury (1970) *op.cit*; Collins (1990) *op.cit.* and Howell, P. Lock, M and Cobb, S. (1988) *The Jonglei Canal: impact and opportunity*, Cambridge University Press.

46. Hays, S.M. (1959) *Conservation and the Gospel of Efficiency*, Harvard University Press, New Haven.

47. Finer, H. (1944) *The TVA: lessons for international application*, ILO, Montréal.

48. Street, E. (1981) 'The role of electricity in the Tennessee Valley Authority', pp. 233–252 in S.K. Saha and C.J. Barrow (eds) *River Basin Planning: theory and practice*, Wiley, Chichester; see also W. Chandler (1984) *The myth of TVA*, Ballinger, Boston.

49. Moris, J. (1987) 'Irrigation as a privileged solution in African development', *Development Policy Review* 5:99–123.

50. Are, L., Fatokun, J.O. and Togun, S. (1982) 'Experiences in river basin development with special reference to the Ogun-Oshun River Basin Development Authority', *Proceedings of the 4th Afro-Asian Regional Meeting of the International Commission on Irrigation and Drainage*, Lagos Nigeria, Volume 1 (Publ.06): 99–100.

51. Siann, J.M. 'Conflicting interest in river-basin planning: a Nigerian case-study', pp. 215–231 in S.K. Saha and C.J. Barrow (eds) *River Basin Planning: theory and practice*, Wiley, Chichester, p. 231.

52. Salau, A.T. (1986) 'River basin planning as a strategy, for rural development in Nigeria', *Journal of Rural Studies* 2 (4): 321–5 , p. 334.

53. Mustapha, S. (1982) 'Water resource development in Nigeria and prospects of drought', *New Nigerian*, 28th January 1982, p. 1.

54. Faniran, A. (1980) 'On the definition of planning regions: the case for river basins in developing countries', *Singapore Journal of Tropical Geography* 1:9–16.

55. Salau, A.T. (1986) *op. cit.*, p. 334.

56. Scudder, T. (1988) *The African Experience with River Basin Development*, Clark University and Institute for Development Anthropology (to be published by Westview Press); see also T. Scudder (1989) 'The African experience with river basin development', *Natural Resources Forum*, May 1989: 139–148; P.J. Dugan (1990) *Wetland Conservation: a review of current issues and required action*, IUCN, Gland, Switzerland.

57. Leopold, A. to Wildlife Ecology Students at the University of Wisconsin (quoted p. 259 in R. Nash (1983) *Wilderness and the American Mind*, Yale University Press, New Haven (3rd edition).

## Chapter six: Binding the rivers

1. Allen-Williams, D.J. (1975) 'Dams made by men: an engineer's thoughts on attempts to control nature', pp. 427–435 in N.F. Stanley and M.P. Alpers (eds) *Man-made Lakes and Human Health*, Academic Press, London.

2. Adams, W.M. (in prep.) 'Development's deaf ear: downstream users and planned releases from the Bakolori project, Nigeria'.

3. Adams, W.M. (1985) 'River control in West Africa', p. 177–228 in A.T. Grove (ed.) *The Niger and its Neighbours: environment, history, hydrobiology, human use and health hazards of the major West African Rivers*, Balkema, Rotterdam.
4. Dafalla, H. (1975) *The Nubian Exodus*, C. Hurst, London, p.89.
5. Dafalla, H. *op. cit.*, p. 279.
6. Scudder, T. and Colson, E. (1982) 'From welfare to development: a conceptual framework for the analysis of dislocated people', pp. 267–287 in Hansen, A. and Oliver-Smith, A. (eds) *Involuntary Migration and resettlement: the problems and responses of dislocated people*, Westview Press, Boulder, Co.
7. Scudder, T. and Habarad, J. (1991) 'Local responses to involuntary relocation and development in the Zambian portion of the Middle Zambezi Valley', pp. 178–205 in J.A. Mollet (ed.) (1991) *Migrants in Agricultural Development*, Macmillan.
8. Adams, W.M. (1988) 'Rural protest, land policy and the planning process on the Bakolori Project, Nigeria', *Africa* 58: 315–336.
9. The general issues are discussed in W.M. Adams (1990) *Green Development: environmental and sustainability in the Third World*, Routledge, London.
10. Baxter, R.M. (1977) 'Environmental effects of dams and impoundments', *Annual Review of Ecology and Systematics* 8: 255–284; R.L. Welcomme (1970) *Fisheries Ecology of Floodplain Rivers*, Longman, London; R.H. Lowe-McConnell (1985) 'The biology of the river systems with particular reference to the fishes', pp.101–140 in A.T. Grove (ed.) *The Niger and its Neighbours: environment, history, hydrobiology, human use and health hazards of the major West African Rivers*, Balkema, Rotterdam.
11. Welcomme, R.L. (1979) *Fisheries Ecology of Floodplain Rivers*, Longman, London.
12. Davies, B.R. (1979) 'Stream regulation in Africa', pp. 113–142 in J.V. Ward and J.A. Stanford (eds) *The Ecology of Regulated Streams*, Plenum Press, New York.
13. Greenpeace (1991) *Okavango: delta or desert? A question of water*, Greenpeace, London.
14. Drijver, C.A. and M. Marchand (1985) *Taming the Floods: environmental aspects of floodplain development in Africa*, Centre for Environmental Studies, University of Leiden.
15. Drijver, C.A. and Marchand, M. (1985) *ibid.*
16. Drijver, C.A. and Marchand, M. (1985) *ibid.*, Appendix c,p. ll.
17. Tchamba, M.N., Drijver, C.A. and Njiforti, H. (1992) 'The impact of flood reduction on the Waza Logone Region and especially the Waza National Park, Cameroon', Paper to Workshop on *Protected*

*Areas and the Hydrological Cycle*, 4th World Congress on National Parks and Protected Areas, Caracas, Venezuela, February 1992.

18. Hughes, F.M.R. (1984) 'A comment on the impact of development schemes on the floodplain forests of the Tana River of Kenya', *Geographical Journal* 150:230–244.

19. Adams, W.M. (1989) 'Dam construction and the degradation of floodplain forest on the Turkwel River, Kenya', *Land Degradation and Rehabilitation* 1: 1889–1984.

20. Moorehead, R. (1988) 'Access to resources in the Niger Inland Delta in Mali', pp. 27–40 in J.A. Seeley and W.M. Adams (eds) *Environmental Issues in African Development Planning*, Cambridge African Monographs No. 9., Africa Studies Centre, Cambridge.

21. Howell, P., Lock, M. and Cobb, S. (eds) (1988) *The Jonglei Canal: impact and opportunity*, Cambridge University Press, Cambridge.

22. Bingham, M.G. (1982) 'The livestock potential of the Kafue Flats', pp. 95–103 in G.W. Howard and G.J. Williams (eds.) *Proceedings of the National Seminar on Environment and Change: the Consequences of the Hydroelectric Power Development on the Utilization of the Kafue Flats, Lusaka, April 1978*, Kafue Basin Research Committee, University of Zambia, Lusaka.

23. Welcomme, R.L. (1970) *Fisheries Ecology of Floodplain Rivers*, Longman, London.

24. Balasubrahmanyam, S. and Abou-Zeid, S.M. (1982) 'The Kafue River Hydroelectric development', pp. 31–33 in G.W. Howard and G.J. Williams (eds) *Proceedings of the National Seminar on Environment and Change: the Consequences of the Hydroelectric Power Development on the Utilization of the Kafue Flats, Lusaka, April 1978*, Kafue Basin Research Committee, University of Zambia, Lusaka.

25. Balasubrahmanyam, S. and Abou-Zeid, S.M. (1982) 'Post-Itezhitezhi flow pattern of the Kafue and Kafue Flats Region', River Hydroelectric development', pp. 636–67 in G.W. Howard and G.J. Williams (eds) *ibid.*

26. Turner, B.L. (1984) 'The effect of dam construction on the flooding of the Kafue Flats', pp. 1–9 in W.L. Handlos and G.J. Williams (eds) *Development on the Kafue Flats: the Last Five Years*, Kafue Basin Research Committee, University of Zambia, Lusaka.

27. Rennie, J.K. (1982) 'Traditional society and modern developments in Namwala District', pp. 35–45 in G.W. Howard and G.J. Williams (eds) *op. cit.*.

28. Howell, P.P. (1983) 'The impact of the Jonglei Canal in the Sudan', *Geographical Journal* 149:33-48 and P.P. Howell, M. Lock and S. Cobb (eds) (1988) *The Jonglei Canal: impact and opportunity*, Cambridge University Press, Cambridge.

29. The best description of the political background to the Jonglei Canal is given by R.O. Collins (1990) *The Waters of the Nile: hydropolitics and the Jonglei Canal, 1900–1988*, Clarendon Press, Oxford.
30. Drijver, A. and Marchand, M. (1985) *Taming the Floods: environmental aspects of floodplain development in Africa*, Centre for Environmental Studies, University of Leiden (Appendix E) (tiang: *Damalicscus lunatus*).
31. The work, and its extensive results, are discussed in Howell *et al.* (1988), *op. cit.*
32. Howell *et al.* (1988), *op.cit.*, p. 382.
33. Scudder, T. (1990) 'Victims of development revisited: the political costs of river basin development', *Development Anthropology Network* 8(1): 1–5.
34. Khogali, M.M. (1982) 'The problem of siltation in the Khasm el Girba Reservoir: its implications and suggested solutions', pp. 96–106 in H.G. Mensching (ed.) *Problems of the Management of Irrigated Areas of Traditional and Modern Cultivation*, International Geographical Union Working group on resource Management in Drylands, Hamburg.
35. Drijver, C.A. and Marchand, M. (1985) *Taming the Floods: environmental aspects of floodplain development in Africa*, Centre for Environmental Studies, University of Leiden.
36. Adams, W.M. (1985) 'The downsteam impacts of dam construction: a case study from Nigeria', *Transactions of the Institute of British Geographers N.S.* 10:292–302.
37. Kimmage, K. and Adams W.M. (1990) 'Small-scale farmer-managed irrigation in northern Nigeria', *Geoforum* 21:435–443.
38. Lowe-McConnell, R.H. (1985) 'The biology of the river systems with particular reference to the fishes', pp. 101–140 in A.T. Grove (ed.) *The Niger and its Neighbours: environment, history, hydrobiology, human use and health hazards of the major West African Rivers*, Balkema, Rotterdam.
39. Chisholm, N.G. and Grove, J.M. (1985) 'The Lower Volta', pp. 229–250 in A.T. Grove (ed.) *The Niger and its Neighbours: environment, history, hydrobiology, human use and health hazards of the major West African Rivers*, Balkema, Rotterdam.
40. IUCN Wetlands Programme Newsletter No.2, November 1990.
41. Drijver, C.A. and Marchand, M. (1985) *op. cit.*.
42. Welcomme, R.L. (1970) *Fisheries Ecology of Floodplain Rivers*, Longman, London.

43. Horowitz, M.M. (1989) 'Victims of development' *Development Anthropology Network* 9(1): 8–15 and in *Haramata* (IIED Drylands Programme) 14 (December 1991)

44. Horowitz, M.M. (1989) 'Victims of development' *Development Anthropology Network* 7(2): 1–8 (page 5).

45. Adams, W. M. and Hollis, G.E. (1989) *Hydrology and Sustainable resource development of a Sahelian Floodplain Wetland*, Hadejia-Nguru Wetlands Conservation Project, London.

46. Kimmage, K. (1990) 'Nigeria's home-grown dust-bowl', *New Scientist*, 7 July 1990: 42–4 and K. Kimmage (1991) 'Small-scale irrigation initiatives in Nigeria: the problems of equity and sustainability', *Applied Geography* 11: 5–20; K. Kimmage (1991) 'The evolution of the "Wheat trap": the Nigerian wheat boom', *Africa* 60:471–500.

47. Examples of conflicts over the resources of smaller wetlands in drylands are given by I. Scoones (1991) 'Wetlands in drylands: key resources for agricultural and pastoral production in Africa', *Ambio* 20:366–371.

48. Drijver, C.A. and Marchand, M. (1985) *op. cit.*.

49. The background to CBA and EIA is discussed in W.M. Adams (1990) *Green Development: environment and sustainability in the Third World*, Routledge, London.

50. O'Riordan, T. (1990) 'Major Projects and the Environment: 1. on the greening of major projects', *Geographical Journal* 156:141–148.

51. Goodland, R. (1990) 'Major projects and the Environment: 2. environment and development: progress of the World Bank', *Geographical Journal* 156: 149–157.

52. Goodland, R. (1989) 'The World Bank's New Policy on the environmental aspects of dam and reservoir projects', Energy-Environment-Development 1 *(World Bank Reprint Series No. 458)*.

## Chapter seven: Watering the savanna

1. Rydzewski, J.R. (1987) 'Irrigation project appraisal', pp.201–227 in J.R. Rydzewski (ed.) *Irrigation Development Planning*, Wiley, Chichester.

2. Hart, K. (1982) *The Political Economy of West African Agriculture*, Cambridge University Press, Cambridge.

3. Stern, P.H. (1979) *Small Scale Irrigation: a manual of low-cost water technology*, Intermediate Technology Publications Ltd, London, p. 22.

4. World Bank (1990) *Sub-Saharan Africa: from crisis to sustainable growth*, World Bank, Washington, p.89–90.
5. Lipton, M. and Longhurst, R. (1989) *New Seeds and Poor People*, Johns Hopkins University Press, Baltimore.
6. Lipton, M. and Longhurst, R. (1989) *op.cit.*, p.380.
7. Lipton, M. and Longhurst, R. (1989) *op.cit.*, p.70.
8. Lipton, M. and Longhurst, R. (1989) *op.cit.*, p.71.
9. Lipton, M. and Longhurst, R. (1989) *op.cit.*, p.359.
10. FAO Investment Centre (1986) *Irrigation in Africa South of the Sahara*, Food and Agriculture Organisation Investment Centre Technical Paper No.5, Rome.
11. World Bank (1990) *Sub-Saharan Africa: from crisis to sustainable growth*, World Bank, Washington, Table 8.
12. Carruthers, I. (1978) 'Contentious issues in planning irrigation schemes' *Water Supply and Management* 2:301–308.
13. Adams, W.M. (1990) 'How beautiful is small? scale, control and success in Kenyan irrigation', *World Development* 18: 1309–1323.
14. Salau, A.T. (1985) 'River basin planning as a strategy for rural development in Nigeria' *Journal of Rural Studies* 2: 321–335.
15. Carter, R.C., Carr, M.K.V. and Kay, M.G. (1983) 'Policies and prospects in Nigerian irrigation', *Outlook on Agriculture* 12:73–76.
16. Palmer-Jones, R.W. (1987) 'Irrigation and the politics of development in Nigeria', pp. 138–167 in M. Watts (ed.) *State, Oil and Agriculture in Nigeria*, University of California Press, Berkeley.
17. Nwa, E.U. and Martins, B. (1982) 'Irrigation development in Nigeria', in *Proceedings of the Fourth Afro-Asian Regional Meeting of the International Commission on Irrigation and Drainage*, Lagos, Nigeria, Vol. II (Lagos: ICID), pp. 1–18.
18. Clarke, C.L. and Anderson, J. (1987) 'The feasibility report', pp.239–260 in J.R. Rydzewski (ed.) *Irrigation Development Planning*, Wiley, Chichester.
19. Carruthers, I (1981) 'Neglect of O & M in irrigation: the need for new sources and forms of support', *Water Supply and Management* 5: 53–65.
20. World Bank (1990) *Sub-Saharan Africa: from crisis to sustainable growth*, World Bank, Washington, p.27.
21. Kolawole, A. (1987) 'Environmental change and the South Chad Irrigation Project (Nigeria)', *Journal of Arid Environments* 13:169–176.
22. Adams, W.M. (1990) 'How beautiful is small? scale, control and success in Kenyan irrigation', *World Development* 18:1309–1323.
23. World Bank (1985) *Bura Irrigation Settlement Project Mid-Term Evaluation Report 1984. Main Report*, World Bank, Washington, p.16.

24. *The Standard* (1986) 'Bura: watever happened?', Nairobi, 23 January 1986.

25. Standard procedures for feasibility studies and project appraisal are described in C.L. Clarke and J. Anderson (1987) 'The feasibility report', pp.239–260 in J.R. Rydzewski (ed.) *Irrigation Development Planning*, Wiley, Chichester; and in J.R. Rydzewski (1987) 'Irrigation project appraisal',pp.201-227 in J.R. Rydzewski (ed.) *Irrigation Development Planning*, Wiley, Southampton.

26. Chambers, R. and Moris, J. (eds) (1973) *Mwea: an integrated rice settlement in Kenya*, Weltforum Verlag, Munich.

27. Etuk, E.G. and Abalu, G.O.I. (1982) 'River basin development in northern Nigeria: a case study of the Bakolori Project', in *Proceedings of the Fourth Afro-Asian Regional Meeting of the International Commission on Irrigation and Drainage, Lagos, Nigeria*, Vol. II (Lagos: ICID), pp. 335–346.

28. Annual Report of the Hadejia-Jama'are River Basin Development Authority, 1986.

29. Nwa, E.U. (1982) 'Groundwater problems of the Kano River Irrigation Project of Nigeria', in *Proceedings of the Fourth Afro-Asian Regional Meeting of the International Commission on Irrigation and Drainage, Lagos, Nigeria*, Vol. II (Lagos: ICID), pp.123–137.

30. UNESCO (Man and the Biosphere) (1977) *Development of Arid and Semi-arid Lands: obstacles and prospects*, MAB Technical Notes 6, UNESCO, Paris.

31. Chambers, R. (1989) *Managing Canal Irrigation*, Cambridge University Press, Cambridge.

32. Ghatak, S. and Turner, R.K. (1978) 'Pesticide use in less developed countries: economic and environmental considerations', *Food Policy* 3: 136–146.

33. Bull, D. (1982) *A Growing Problem: pesticides in the Third World*, Oxfam Books, Oxford.

34. Andrae, G. and Beckman, B. (1985) *The Wheat Trap: bread and underdevelopment in Nigeria*, Zed Press, London.

35. Carney, J. and Watts, M. (1990) 'Manufacturing dissent: work, gender and the politics of meaning in a peasant society', *Africa* 60:207–230.

36. Carney, J. and Watts, M. (1990) *op. cit.*

37. Adams, W.M. (1984) 'Irrigation as hazard: farmers' responses to the introduction of irrigation in Northern Nigeria' pp.121–130 in Adams W.M. and A.T. Grove (eds) *Irrigation in Africa: problems and problem-solving*, Cambridge African Monograph No.3.

38. Baba, J.M. (1989) 'The problems of inequality on the Kano River Irrigation Project, Nigeria' pp.140–155 in K. Swindell, J.M. Baba

and M.J. Mortimore (eds) *Inequality and Development: case studies from the Third World*, Macmillan, London.

39. Jega, A. (1987) 'The state, agrarian change and land administration in the Bakolori Irrigation Project', in M. Mortimore, F.A. Olofin, R.A. Cline-Cole and A. Abdulkadir (eds) *Perspectives on land administration and development in northern Nigeria*, (Department of Geography, Bayero University, Kano); K. Swindell and A.B. Mamman (1990) 'Land expropriation in the Sokoto periphery NW Nigeria 1976–86', *Africa* 60(2): 175–187.

40. Adams, W. M. (1988) 'Rural protest, land policy and the planning process on the Bakolori Project, Nigeria', *Africa* 58:315–336.

41. Vainio-Mattila, A. (1987) *Domestic Fuel Economy*, Bura Fuelwood Project, Institute of Development Studies Report 13, University of Helsinki.

42. Saha, S.K. (1982) 'Irrigation planning in the Tana River Basin of Kenya', *Water Supply and Management* 6:261–279.

43. Chimbari, M., Chitsiko, R.J., Bolton, P. and Thompson, A.J. (1991) 'Design and operation of a small irrigation project in Zimbabwe to minimise schistosomiasis transmission', *Overseas Development Institute Irrigation Management Network (African Edition), Paper 9*, October 1991.

44. Caufield, C. (1984) 'Pesticides: exporting death', *New Scientist*, 16 August, pp. 15–17.

45. Schumacher, E. (1983) *Small is Beautiful: economics as if people mattered*, Harper and Row, London.

46. UNESCO (MAB) (1977) *Development of Arid and Semi-Arid Lands: obstacles and prospects*, Man and the Biosphere Technical Notes 6, UNESCO, Paris', p. 18.

47. This account draws extensively on the work of Richard Hogg: R. Hogg (1983) 'Irrigation agriculture and pastoral development: a lesson from Kenya', *Development and Change* 14:577–591; R. Hogg (1987) 'Settlement, pastoralism and the commons: the ideology and practice of irrigation development in northern Kenya', pp. 293–306 in D.M. Anderson and R.H. Grove (eds) *Conservation in Africa: people, policies and practice*, Cambridge University Press, Cambridge.

48. Adams, W.M. (1990) 'How beautiful is small? scale, control and success in Kenyan Irrigation', *World Development* 18: 1309–1323.

49. Overseas Development Institute (1983) *Turkana District development Strategy and Programme 1985–86, 1987–88*, ODI for the Kenyan Ministry of Energy and Rural Development, under assignment for the European Community.

50. Ton, K. and de Jong, K. (1990) 'Pump and flood irrigation: complements for sustainable development in Northern Mali', *Water Resources Development* 6: 122–128.

51. Barrett, H. and Browne, A. (1991) 'Environmental and economic sustainability: women's horticultural production in the Gambia', *Geography* 76:241–248.

52. Adams, W.M. (1990) *op. cit.*

53. Boutillier, J.L. and Schmitz, J. (1987) 'Gestion traditionelle des terres (système de décrue/système pluvial) et transition vers l'irrigation', *Cah. Sci. Hum.* 23:533–554.

54. This process is described in P. Woodhouse and I. Ndiaye (1991) 'Structural adjustment and irrigated agriculture in Senegal', *Overseas Development Institute Irrigation Management Network (African Edition), Paper 7*, October 1991.

55. Niasse, M. (1990) 'Village irrigated perimeters at Doumga Rindiaw, Senegal', *Development Anthropology Network* 8(1):6–11.

56. For example G. Diemer, B. Fall and F.P. Huibers (1991) 'Promoting a smallholder-centred approach to irrigation: lessons from village irrigation schemes in the Senegal River Valley', *Overseas Development Institute Irrigation Management Network (African Edition), Paper 6*, October 1991. They argue that the success of the village irrigation schemes in Senegal is demonstrated by their rapid spread, and that this success is due to the fact that they are 'generated from within local societies' (p.4), and have 'conformity with smallfarmer realities' (p.20).

57. See, for example, A.K. Biswas (1988) 'Irrigation in Africa', *Land Use Policy 3*: 329–385, and J. Moris and D.J. Thom (1987) *African Irrigation Overview: main report*, Water Management Synthesis II Project Report 37, Utah State University.

58. Kimani, J.K. (1984) 'Smallholder irrigation schemes: the Kenyan experience', pp. 259–271 in M.J. Blackie (ed.) *African Regional Symposium on Small Holder Irrigation*, Hydraulics Research Ltd, Wallingford.

59. World Bank (1990) *Sub-Saharan Africa: from crisis to sustainable growth*, World Bank, Washington, p.90.

60. World Bank (1990) *ibid.*, p. 90.

61. World Bank (1990) *ibid.*, p. 93.

62. World Bank (1990) *ibid.*, p. 93.

63. World Bank (1990) *ibid.*, p. 91.

# Chapter 8: Water, people and planning

1. Abraham, R.C. *Dictionary of the Hausa Language*, Crown Agents for the Colonies, London.
2. Irrigation in Sonjo is described by R. Gray (1963) *The Sonjo of Tanzania: an anthropological study of an irrigation-based society*, Oxford University Press for the International African Institute, London; and by T. Potkanski (1987) 'The Sonjo community in the face of change', *Hemispheres* 4: 192–222.
3. Adams, W.M. and Carter, R. (1987) 'Small Scale irrigation in sub-Saharan Africa', *Progress in Physical Geography* 11: 1–27; W.M. Adams (1989) 'Definition and development in African indigenous irrigation', *Azania* 24: 21–27.
4. Ton, K. and de Jong, K. (1990) 'Pump and flood irrigation: complements for sustainable development in northern Mali', *Water Resources Development* 6(2): 122–128.
5. Fadama cultivation in Nigeria is described in Chapter 4, and in B. Turner (1984) 'Changing land-use patterns in the fadamas of northern Nigeria', pp.140–170 in E.P. Scott (ed.) *Life Before the Drought*, Allen and Unwin, Hemel Hempstead. The introduction of petrol pumps is discussed in K. Kimmage and W.M. Adams (1990) 'Small-scale farmer-managed irrigation in northern Nigeria', *Geoforum* 21 (4):435–443.
6. Kimmage, K. and Adams, W.M. (1990) 'Small-scale farmer-managed irrigation in northern Nigeria', *Geoforum* 21: 435–443.
7. Kenna, J. and Gillet, B. (1985) *Solar Water Pumping: a handbook*, Intermediate Technology Publications.
8. There are a number of good reviews outlining the importance of small-scale irrigation in Africa, notably H. Underhill (1984) *Small-scale Irrigation in Africa in the Context of Rural Development*, FAO, Rome; J. Moris, D.J. Thom and R. Norman (1984) *Prospects for Small-scale Irrigation in the Sahel*, Water Management Synthesis II Report, Utah State University, and J. Moris and D.J. Thom (1987) *African Irrigation Overview: main report*, Water Management Synthesis II Report, Utah State University; J.R. Moris and D.J. Thom (1990) *Irrigation Development in Africa: lessons of experience*, Westview Press, Boulder, Co.
9. FAO (1986) *Irrigation in Africa South of the Sahara*, FAO Investment Centre Technical Paper No.5, Rome.
10. Moore, M. (1988) 'Maintenance before management: a new strategy for small-scale irrigation tanks in Sri Lanka', *ODI/IIMI Irrigation Management Network Paper 88/2e*.

11. Vermillion, D.L. (1989) 'Second approximations: unplanned farmer contributions to irrigation design', *ODI/IIMI Irrigation Management Network Paper 89/2c.*

12. Kimmage, K. (1990) 'Nigeria's home-grown dust bowl', *New Scientist* 127 (1724): pp. 42–44; and K. Kimmage, (1991) 'Small scale irrigation initiatives in Nigeria: the problems of equity and sustainability', *Applied Geography* 11: 5–20; K. Kimmage (1991) 'The evolution of the "Wheat trap": the Nigerian wheat boom', *Africa* 60:471–500.

13. Adams, W.M. (1987) 'Approaches to water resource development, Sokoto Valley, Nigeria: the problem of sustainability', pp.307–325 in D.M. Anderson and R.H. Grove (eds) *Conservation in Africa: people, policies and practice*, Cambridge University Press, Cambridge.

14. Adams, W.M. (1991) 'Large scale irrigation in northern Nigeria: performance and ideology', *Transactions of the Institute of British Geographers* N.S. 16: 287–300.

15. FAO (1986) *op.cit.*

16. Speelman, J.J. (1990) 'Designs for sustainable farmer-managed irrigation schemes in Sub-Saharan Africa: a compilation of results of recent international meetings', *ODI/IIMI Irrigation Management Network Paper* 90/lf.

17. Speelman, J.J. (1990) *op. cit.*, p. 11.

18. Barrett, H. and Browne, A. (1991) 'Environmental and economic sustainability: women's horticultural production in the Gambia', *Geography* 76: 241–248.

19. Diemer, G., Fall, B. and Huibers, F.P. (1991) 'Promoting a smallholder-centred approach to irrigation: lessons from village irrigation schemes in the Senegal Valley', *Overseas Development Institute Irrigation Management Network (African Edition) Paper 6*, October 1991.

20. A contrast is provided by the summary of current good practice by J.R. Rydzewski (ed.) (1987) *Irrigation Development Planning: an introduction for engineer's*, Wiley, Southampton.

21. Moris, J. and Thom, D.J. (1987) *African Irrigation Overview: main report*, Water Management Synthesis Project 37, Utah State University.

22. Speelman, J.J. (1990) *op. cit.* p. 5.

23. I have taken a pragmatic line in this book, because I believe that it is logical and that it may be effective in changing the way the development of Africa's rivers is approached. Apart from an intuitive belief in the power of rational debate and consensus, I am also influenced by a feeling that as an outsider in debates about Africa, an important role is to interpret and influence the ideas and atti-

tudes of those other outsiders who have the financial, political and technical power to transform its environment and the lives of its people. This is a paternalistic agenda, but I think that it is an honest one. I believe that change requires dialogue, and that there is little point in starting a dialogue with unreasonable words and a closed mind. Others will probably argue for a more confrontational stance. At times such an attitude to imposed development is very compelling.

24. Scudder, T. (1980) 'River-basin development and local initiative in African savanna environments', pp. 383–405 in D.R. Harris (ed.) *Human Ecology of Savanna Environments*, Academic Press, London; T. Scudder (1989) 'The African experience with river basin development', *Natural Resources Forum* May 1989: 139–148; T. Scudder (1991) 'The need and justification for maintaining transboundary flood regimes: the Africa case', *Natural Resources Journal* 31(1): 75–107; T. Scudder (in press) *The African Experience with River Basin Planning*, Westview Press, Boulder, Co..

25. Scudder, T. (1991) *op. cit.*

26. Scudder, T. (1991) *op. cit.* p. 19.

27. White, R.R. (1990) 'Agricultural policy options in the face of uncertainty: the case of Senegal', pp. 151–163 in E.A. McDougal (ed.) *Sustainable Agriculture in Africa*, Africa World Press Inc., Trenton, New Jersey.

28. This work is reviewed in M. Salem-Murdock and M.M. Horowitz (1991) 'Monitoring development in the Senegal river basin', *Development Anthropology Network* 9(1): 8–15.

29. Salem-Murdock and Horowitz *op.cit.*, p. 9.

30. Niasse, M. (1990) 'Village Irrigated Perimeters at Doumga Rindiaw, Senegal', *Development Anthropology Network* 8(1):6–11.

31. Hollis, G.E. (1990) *Senegal River Basin Monitoring Activity, Hydrological Issues Part II*, Institute for Development Anthropology, Binghampton NY.

32. Salem-Murdock and Horowitz *op.cit.*, p. 13.

33. This complex issue has been tackled for the River Senegal by G.E. Hollis (1990) *op. cit.*

34. McHarg, I.L. (1969) *Design with Nature*, Natural History Press, Garden City NY, for the American Museum of Natural History.

35. The background to and arguments of the World Conservation Strategy are discussed in W.M. Adams (1990) *Green Development: environment and sustainability in the Third World*, Routledge, London.

36. Hollis, G.E. (1990) 'Environmental impacts of development on wetlands in arid and semi-arid lands', *Hydrological Sciences Journal* 35: 411–428.

37. Dugan, P.J. (ed.) (1990) *Wetland Conservation: a review of current issues and required action*, International Union for the Conservation of Nature and Natural Resources, Gland, Switzerland.

38. Pirot, J.Y. (1990) 'Montreux Workshops', *IUCN Wetlands Programme Newsletter* 2:13, November 1990.

39. Toulmin, C. and Chambers, R. (1990) *Farmer-First: achieving sustainable dryland development in Africa*, IIED Drylands Network Programme Paper 19, June 1990.

40. McRobie, G. (1981) *Small is Possible*, Abacus Books, London, p.1–2.

41. R. Chambers (1991) 'Sustainable rural livelihoods: a key strategy for people, environment and development', Unpublished manuscript of paper to IIED Conference on Sustainable Development, London, 28–30 April 1987, p.14.

42. Danida (1988) *A Strategy for Agriculture, Livestock Husbandry and Forestry in Arid and Semi-Arid Areas: a plan of action for the integration of environmental considerations into Danish development assistance*, Danida, Copenhagen (Parts I and II by M.E. Adams).

43. Chambers, R. (1991) 'In search of professionalism, bureaucracy and sustainable livelihoods for the 21st Century', *IDS Bulletin* 22(4): 5–11, p. 5.

44. Chambers, R. (1991) *op.cit.*, p. 7.

45. Chambers, R. (1991) *op.cit.*, p. 8. see also R. Chambers (1983) *Rural Development: putting the last first*, Macmillan, London.

46. Chambers, R. (1989) 'Reversals, institutions and change', pp. 181–195 in R. Chambers, A. Pacey and L.A. Thrupp (eds) *Farmer First: farmer innovation and agricultural research*, IT Publications, London.

47. Chambers, R. (1991) *op.cit.*

48. Dugan, P.J. (1991) Wetlands, regional planning and the development assistance community', *Landscape and Urban Planning* 20:211–214.

49. Turner, K. (1991) 'Economics and wetland management', *Ambio* 20:59–63.

50. Turner, K. (1991) *op.cit.* and J.P. Hunter (1991) 'Economics of wetlands in southern Africa: a search for a justification for their conservation', *Splash* 7(2):16–20.

51. Bebbington, A. (in press) 'Peasant knowledge, peasant agency, peasant organisation: theory and relevance in indigenous agriculture', in D. Booth (ed.) *New Directions in Social Development Research: Relevance, Realism and Choice* (in press). Populist approaches to rural change are criticized in M. Watts (1989) 'Popu-

lism and the politics of land use' *African Studies Review* 26: 73–83. The importance of securing rights and gains is stressed by R. Chambers (1991) 'Sustainable rural livelihoods: a key strategy for people, environment and development', Unpublished manuscript. of paper to IIED Conference on Sustainable Development, London, 28–30 April 1987, p.14.

# Index

Printed and bound by CPI Group (UK) Ltd, Croydon, CR0 4YY

01/11/2024

01782615-0002